Killer Algae

Killer Algae

ALEXANDRE MEINESZ

TRANSLATED BY DANIEL SIMBERLOFF

WITH A FOREWORD BY DAVID QUAMMEN

The University of Chicago Press *Chicago and London*

ALEXANDRE MEINESZ is professor of biology at the University of Nice and is a member of various scientific committees on the environment, including the National Geographic Institute. He is currently the president of the Environmental Commission for the regions of Provence, the Alps, and the Côte d' Azur.

DANIEL SIMBERLOFF is the Nancy Gore Hunger Professor of Environmental Studies at the University of Tennessee, Knoxville. He is the editor of *Ecological Communities* and the author of *Strangers in Paradise: Impact and Management of Nonindigenous Species in Florida.*

The University of Chicago Press, Chicago 60637
The University of Chicago Press, Ltd., London
© 1999 by The University of Chicago
Foreword © 1999 by David Quammen
All rights reserved. Published 1999

08 07 06 05 04 03 02 01 00 99   5 4 3 2 1

ISBN (cloth): 0-226-51922-8

Originally published as *Le roman noir de l'algue "tueuse": Caulerpa taxifolia contre la Méditerranée,* © Éditions Belin, 1997

Library of Congress Cataloging-in-Publication Data

Meinesz, Alexandre
    [Roman noir de l'algue "tueuse". English]
    Killer algae / Alexandre Meinesz ; translated by Daniel Simberloff ; with a foreword by David Quammen.
        p.   cm.
    Includes bibliographical references and index.
    ISBN 0-226-51922-8 (cloth : alk. paper)
    1. Caulerpa taxifolia—Mediterranean Sea.   2. Toxic marine algae—Mediterranean Sea.   3. Alien plants—Mediterranean Sea.
    I. Simberloff, Daniel.   II. Title.
QK569.C37M4513   1999
579.8'35—dc21
                                                                99-32502
                                                                  CIP

This book is printed on acid-free paper.

Contents

Foreword

Back in 1958, with great foresight and little fanfare, a British ecologist named Charles Elton published a book entitled *The Ecology of Invasions by Animals and Plants*. This was several years before Rachel Carson's *Silent Spring* delivered its grim message about chemical pesticides, a decade before the ticking of Paul Ehrlich's *The Population Bomb* became audible, a dozen years before the first Earth Day, and a generation before any widespread public concern about rainforest destruction, ozone depletion, global climate change, or the wholesale extinction of species resulting from all manner of human impact upon the world's landscapes. Elton's book went largely unnoticed amid this dawning awareness of ecological tribulation. It was a small volume, engagingly written, that discussed such matters as how the American muskrat colonized Europe.

It told the story of Asian chestnut blight in New England forests. It traced the dispersal of sea lampreys up the Welland Canal into Lake Erie, the transport of malaria-carrying *Anopheles* mosquitoes aboard boats crossing from Africa to Brazil, and the inadvertent introduction of an American species of limpet into English oyster beds. It passingly mentioned *Asparagopsis armata*, a seaweed (more precisely, a red alga) of Australian provenance that, around 1923, began showing up along the Atlantic coast of France. It alerted readers that bubonic plague was alive and well in the gerbils of South Africa and in the ground squirrels of California. It described how the giant African snail got to Hawaii. It charted the pestiferous abundance, in New Zealand, of European red deer. By way of these and other minatory tales, Elton sketched the whole subject of invasion biology—that is, the dynamics and the consequences of species transfer from one ecosystem to another.

Although sometimes such transfers occur naturally, within recent centuries the rate has increased drastically, due both to intentional and to accidental human actions. In some cases those transfers have resulted in what Elton called "ecological explosions." His choice of the word *explosion* was deliberate, he wrote, "because it means the bursting out from control of forces that were previously held in restraint by other forces."[1] When the transplanted species happens to be an aggressive, opportunistic form of animal or plant, capable of fierce predation or competition against unfamiliar native species within the new ecosystem, and freed suddenly from its own familiar predators, competitors, and pathogens left behind in the ecosystem of origin, the results can indeed be explosive. Elton warned that "we are living in a period of the world's history when the mingling of thousands of kinds of organisms from different parts of the world is setting up terrific dislocations in nature. We are seeing huge changes in the natural population balance of the world."[2] But his warning made no such immediate impact as Carson's or Ehrlich's, partly because exotic seaweeds and peripatetic muskrats failed to capture the horrified attention of many readers.

Environmental poisoning, clearcutting rainforest to make rice fields or plywood, overharvest of fishes or whales, human population growth and its attendant destruction of habitat—these are obvious forms of damage to the biosphere. Ecological mayhem caused by non-native species is more subtle. But history has caught up with Charles Elton's foresight, and *The Ecology of Invasions* can now be seen as one of the central scientific books of our century. Although much has changed since 1958 (including the identity of the planet's currently most menacing species of alga), one point remains constant: Human-mediated invasions are a big, serious, costly problem, destined to get worse and more noticeable in the near future.

Biologists and some astute policy-makers have recognized that fact, though the wake-up among government agencies, scientific institutions, and society at large is only beginning. A study released in 1993 by the congressional Office of Technology Assessment found that about 4,500 exotic species have established populations within the U.S., of which 15 percent cause serious harm; native species have been driven extinct, ecosystems have been destabilized and depauperated, and the economic losses from just 79 select exotics have amounted to $97 billion. The more nefarious invaders include *Melaleuca* trees in the Everglades, fire ants throughout the rural South, salt cedar in the river bottoms of Arizona and New Mexico, red mangrove in the bays of Hawaii, water hyacinth in north Florida rivers, the brown tree snake in Guam, the zebra mussel in the Great Lakes, the Mediterranean fruit fly in California, and many others. A small number of contemporary scientists, in the tradition of Elton, have produced important primary work and broad syntheses on the subject, notable among them being Hal Mooney at Stanford, James Drake at the University of Tennessee, James Carlton at Williams College, Mark Williamson at the University of York in Britain, and Daniel Simberloff, now also at the University of Tennessee. Simberloff's engagement is especially telling, since besides being a brilliant ecologist with a practical grounding in insect-pest management, he is a gimlet-eyed skeptic with a long history of rigorous thinking about scientific aspects of conservation. If the problem seems real and dire to Dan Simberloff, the problem must be very real and very dire.

Simberloff recently wrote that "invasion biology has burgeoned in the last decade into a main focus of conservation biology and ecology. Realization is growing that nonindigenous species are not only a monumental economic threat but an insidious and pervasive conservation problem."[5] That realization has led him to establish an institute on invasion biology at the University in Knoxville. It has also inspired him to seize personally the task of translating Alexandre Meinesz's important book for an American audience.

The story recounted here by Professor Meinesz is unique in its particulars but global in its meanings. I won't dissipate its force by attempting a summary, and I won't presume to tease out the salient lessons that Meinesz himself extracts quite well. I will just offer two thoughts. First, what makes the case of *Caulerpa taxifolia* extraordinary is not the eerie, pestilential proliferation of a tropical alga in the cool waters of the northern Mediterranean, nor the maddeningly squandered opportunity to stop it at an early stage, but the very fact that *this book* exists. Meinesz's fight against *C. taxifolia*—and against the arrogant, ill-informed officials who allowed it to prosper—may be a lost battle, but his combat journal is enormously valuable. We will never have any such well-documented record of how the gypsy moth, the whelk tingle, or the house sparrow established their beachheads and rolled on to conquest. Second, and a point about which I urge you to make no mistake: This is not a little book about some noxious alga. This is a little book, like Charles Elton's, about life on Earth.

David Quammen

Preface

In the early 1980s, the curator of the tropical aquarium at Stuttgart, Germany, noticed the exceptional properties of a beautiful green alga, *Caulerpa taxifolia*, used as decoration in the presentation of multicolored tropical fishes. In contrast to other algae, it does not wither, it grows with astounding vigor, it resists cool water temperatures, and it serves as a secondary food source for herbivorous tropical fishes. Specialists quickly learned about these qualities, and public aquaria acquired cuttings.

This is how it arrived at the Oceanographic Museum of Monaco, where it was cultivated beginning in 1982. Two years later, the alga was discovered in nature, under the windows of this celebrated building. At that time, the beautiful stranger occupied only a square meter of Mediterranean bottom. Six years later, the alga was noted on the French coast five kilometers from Monaco; its detrimental impact on coastal ecosystems was deplored. The alga grows everywhere, from the surface to the lower limits of underwater vegetation. It grows as well in front of capes swept by storms and currents as on the soft bottoms of sheltered bays, on the polluted mud of harbors as on stretches of bottom with a diverse flora and fauna. Highly toxic, it barely interests herbivores; they have not hindered its spread. It is thus growing unrestrained, covering and then eliminating many plant and animal species. A new equilibrium is reached when the alga forms a dense, uniform carpet that persists from year to year.

After having selected it for aquaria from among numerous imported algal species, after having dumped it into the sea, humans fostered its dissemination in nature. Yacht anchors and fishing gear have carried it from anchorage to anchorage and from harbor to harbor, sometimes over great distances. The Italian and Spanish coasts were reached by 1992, that of Croatia by 1995. By late 1997, ninety-nine invaded sites totaling more than 4,600 hectares have been inventoried.

No one has ever been killed by *Caulerpa taxifolia*, known as the "killer alga." For, contrary to what its media nickname might suggest, this prolific alga is primarily an ecological threat. All relevant research indicates an unlimited spread. Its control is more difficult every year, and its eradication, envisaged at the beginning of the invasion, can now be classed only as a utopian dream. The introduction of this dangerous alga therefore threatens to initiate a profound disruption of the coastal Mediterranean environment. The story of the "killer alga" has, unfortunately, just begun.

How did we reach this point? When the first scientific publications confirmed the threat, the alga was, contrary to any reasonable expectation, defended by other scientists who tried to argue that its appearance was a natural event. This was the beginning of a long, fantastic polemic. The affair is even stranger because the place where the alga was introduced to the Mediterranean is truly incongruous—the principality of Monaco, a state with one of the highest standards of living. This is far from the Polish forests devastated by acid rain, far from the Aral Sea dried up by diverting water, far from the third world where overpopulation engenders overexploitation of natural resources. More remarkably, the first signs of the invasion were observed just in front of a palace of the sea, a landmark of marine biology: the Oceanographic Museum of Monaco. This prestigious palace, built between 1899 and 1910 by Prince Albert I of Monaco (a highly erudite and competent oceanographer), was directed from 1957 through 1988 by Commander Jacques-Yves Cousteau, a person emblematic of the sea.

The polemic has been heated, fed by the defense of many different interests. It was able to break out because key scientific and government authorities were lax and because of disdain for a problem that does not directly threaten human health. An abundance of communications on this affair masks inadequate knowledge and a failure of government experts. The object of byzantine debates between scientists, government experts, public figures, and the media, this sterile controversy slowed the recognition of the threat. The threat was long underestimated while the time during which it might have been successfully contained dribbled away. The alga grew inexorably, and it still grows, disturbing the marine environment . . . and the human intellect.

A university researcher who specializes in *Caulerpa*, and a diver devoted to defending marine life, I was the first to sound the alarm in 1989. An actor in or observer of all the intricacies of this affair, I encountered deplorable actions in the face of a concrete threat to biodiversity, at a propitious time in countries in which everything could have been undertaken to allow a rational and rapid mastery of the situation. Having taken it upon myself to alert the authorities and then the media about the imminent threats, I am undertaking in this book to describe the years of battle that I have subsequently lived. This fact explains the personal tone of this report.

Of course, recalling facts and activities will arouse in the reader some remorse and much revulsion and indignation. But this first history of the killer alga is much more than the simple, lively chronicle of a trivial ecological accident. It is also an analysis of the social and political mechanisms that raised obstacles to the successful management of a potentially grave environmental threat. The passions that were unleashed, often extremely heated, were the result of a failure of our institutions to function properly. This account can also be read as an example of the conflicts in scientific and administrative hierarchies with respect to the "affairs" that have shaken up our industrialized countries at the end of

this century and in which scientists very quickly played the role of sentinel; they detected the threat and gave the alarm. But many obstacles
delayed the decisions that would have to have been made to avoid the
dreadful sequels that we now observe. Though the alga is not a killer,
the damage caused to our environment is already manifest, and the ecological consequences for the fauna and flora—and finally for humans
who exploit them—can be very grave. The years of inaction produce
and will continue to produce, in every case, an enormous cost for society.

The first six chapters of this book present, with details and references, the chronology of the events and their socio-political context,
from the introduction of the alga at Monaco through the delayed recognition of its harmful nature. The seventh and last chapter reflects on
the three causes underlying all the incoherence and negligence that
characterized this affair: scorn for biodiversity at the decision-making
level, the decline of the sciences of nature, and the evolution of the ways
in which scientific information is communicated. The seventh chapter
also locates the invasion of *Caulerpa taxifolia* amidst the swelling tide
of exotic species invading terrestrial, freshwater, and marine habitats
all over the earth, a phenomenon that now ranks second among global
threats to biodiversity.

As beautiful as a flower, *Caulerpa taxifolia* still poses many scientific questions. It has aroused passions in people with diverse interests.
Some actors in the black tale of this "natural history" displayed an incomprehensible attitude, lamentable for people of their rank, given the
responsibility that society has given them. The alga has thus become
the sort of evil flower that Baudelaire must have glimpsed when he
wrote his prescient poem:[1]

Both of you are discreet, dim, shadow-ridden:
Man, none has plumbed your soul's abyss; and, sea,
No one has pierced your wealth's dark mystery,
So jealous, you, to keep your treasures hidden!

Acknowledgments

I thank my adoptive family, my in-laws. My father-in-law, Jean Clary, has continued to encourage me; he has tirelessly reread and corrected the manuscript and has steered me to many readings that contributed heavily to the final product. I thank my wife, Claude Clary-Meinesz, and her sister, Monique Lehew, who have spent many nights seeking out errors of syntax or style.

I owe a debt to Michèle Febvre, lecturer in the laboratory of the coastal marine environment, who has worked over and polished the manuscript. I have done my best to follow his advice and to absorb his strong but appreciated criticisms. I must also mention Jean-Michel Cottalorda and Danièle Chiaverini, technicians in my laboratory; they took responsibility for the figures in this book.

André Chapot, André Giordan, Henri Helly, Arlette Midorge, Danièle Pesando, and Michel Simonian, with a variety of backgrounds, were consulted in the final phase of writing, and the contribution of each was precious.

The other members of my laboratory—Jean de Vaugelas, Gilberte Caye, Anne-Marie Campredon and all the students, foreign interns, and scientific conscientious objectors who passed through my laboratory from 1990 to 1996—have participated greatly in efforts to convince the xvi authorities of the problem and in research on *Caulerpa taxifolia* (at certain times the laboratory, target of many attacks, resembled an armed camp). These "Robin Hoods of the Sea"—Jérôme Blachier, Michael Braun, Daphné Buckles, Nelson Carvalho, Patrick Chambet, Stéphane Charrier, Thierry Commeau, Laurent Delahaye, Frederic Jaffrenou, Bruno Hesse, Sandy Ierardi, Teruhisa Komatsu, Rodolphe Lemée, Luisa Mangialajo, Xavier Mari, Jennifer Melnyk, Ulrike Meyer, Heike Molenaar, Gilles Passeron-Seitre, Dzitka Pietkiewicz, Kelly Ryder, Anne Suissa, Thierry Thibault, and Valérie Vidal—have dived or overseen cultures and experiments in the laboratory with passion, tenacity, and perseverance. Taking time out from their studies or their vacations, they have greatly contributed to the demonstration that the alga exists, is spreading, and harms the environment. It was the cohesion of this group of scientific divers, united by a devotion to protecting marine habitats, that overcame so many obstacles and so much indifference or scorn.

I must also thank all those who supported me early in the struggle and who have greatly contributed to the confirmation of our hypotheses. Among these new allies I want to cite Professor Charles-François Boudouresque and the members of his laboratory and of the organization GIS–Posidonie that he heads. They took on the heavy burden of finding and managing the majority of the research funds for the study of *Caulerpa taxifolia.*

Similarly, I am eager to acknowledge "Mr. Caulerpa," André Manche, former director of the Port-Cros National Park, assigned by Ministers of the Environment Michel Barnier and then Corinne Lepage to head the project on the problems posed by the algal spread. He dived to the *Caulerpa* prairies and very quickly became an adherent of the "alarmist" ideas. For all his efforts of persuasion and communication in the highest ministerial circles, unfortunately poorly compensated: thank you!

From the Discovery of the Alga in Monaco to Its Arrival in France

1988–1989: The Discovery in Monaco

In February 1988, a student in the laboratory of marine biology and ecology at the University of Nice found a tropical alga of the genus *Caulerpa*, cultivated in aquaria of the Oceanographic Museum of Monaco, growing in the open sea, just beneath the windows of this palace of marine biology. A museum diver—an aquarium employee—had alerted him to its presence and told him of his astonishment at seeing the alga survive in the Mediterranean. I was then very close to this student; we both had fond memories of dives on the coast of Saudi Arabia, where he had helped me to collect algae of the Red Sea. He knew of my great interest in algae of the order Caulerpales, and he was confident I would be discreet. I agreed to wait quietly for other news of this introduction of an exotic species.

As months passed, I often asked him if the alga could still be found in front of the Museum. What species was it? What did this stand look like? Was it abundant? How had it survived the winter? My curiosity was legitimate—for eleven years I had studied algae of the order Caulerpales, first for my higher education degree in 1969, then for my master's thesis in 1973, and finally for my doctorate in 1980. I had described the biology of algae in the Caulerpales, and I had published the first study of their very unusual reproductive cycle. One might say that the arrival of this *Caulerpa* just several miles from my laboratory was foreordained.

I had begun to study the Caulerpales of tropical waters while accompanying my colleague Jean Jaubert in Polynesia and the Red Sea. Through the sponsorship of Raymond Vaissière (then director of our laboratory as well as assistant director of the Oceanographic Museum of Monaco and the first scientist of the Cousteau team to have lived in an underwater dwelling), I had participated in two expeditions in the American underwater laboratory called the Hydrolab, based in the Virgin Islands in the heart of the Caribbean. Two weeks of living on compressed air (saturating my blood with nitrogen) had allowed me to conduct research while diving, eight hours each day, on the underwater prairies of *Caulerpa* algae in the tropical Atlantic.

On each expedition, I deposited hundreds of dried specimens of Caulerpales in an herbarium and thus familiarized myself with the classification of these algae; their morphology is highly variable and changes with the surrounding environment. Their ecology, studied in the course of thousands of dives, and their reproduction, watched for hours by microscope, held no more secrets for me; I knew them thoroughly. My published observations were as relevant to all *Caulerpa* species—those of Northern Tuamotu (near Tahiti), Papua New Guinea, and the six main species of the Mediterranean.[1]

In 1977, Jean Jaubert and I coauthored an article in an aquarium journal recommending algae of the order Caulerpales as decorations for tropical aquaria. At the beginning of the 1980s, different species of Caulerpales began to grace the tropical aquaria of the Monaco Museum, replacing dreary mineral decorations made of local rocks or dead corals. The interesting traits of *Caulerpa taxifolia* were just beginning to be recognized in professional aquarium circles.

In the spring of 1989, I decided to accompany my museum infor- mant to see firsthand the *Caulerpa* species grown at the Museum. He took me to the office of Dominique Bézard, adjunct of the director of the aquaria and himself a diver, who showed me the alga in stock aquaria as well as completely identical specimens, collected just beneath the Museum. I took several fragments to my laboratory. I held in my hands a *Caulerpa* with highly serrate leaves, feathery, a very striking alga resembling a beautiful green fluorescent fern. As for its identity, a close rereading of many articles and examination of several dozen herbarium specimens left no doubt: it was *Caulerpa taxifolia*. This *Caulerpa* is aptly named, as its fronds (or "leaves," *folia* in Latin) resemble the branches of a conifer, the yew (*Taxus*, in Latin)—it is the *Caulerpa* "with yew leaves."

Until then, the alga *Caulerpa taxifolia* had never been seen in the Mediterranean. It grows naturally in tropical oceans. Among my herbarium specimens, I found samples gathered on the coral reefs of the Caribbean (Guadeloupe and the Virgin Islands) and Polynesia (Moorea, near Tahiti). After gathering several specimens in 1982 near Jiddah, Saudi Arabia, I had noticed that only one other colleague—an Israeli— had encountered it in the Red Sea, and this just ten years previously. The alga was supposed to be very rare in this region! I reread the notes from my tropical expeditions and the underwater scenes returned to me; I had always found *Caulerpa taxifolia* in the form of spindly, small, isolated plants, never dense clumps.

Bézard told me of his surprise at seeing the alga spread little by little below the Museum: there was a lot of it between three and thirty meters deep. How had it reached the Mediterranean? His hypotheses were embarrassing. Fragments thrown out a Museum window? Microscopic eggs drained directly into the sea with aquarium water? Some other pathway of contamination? The secret of its acclimatization to the open marine environment, unmasked by chance by a museum diver, seemed to me a bit inconvenient for the Museum.

At the end of April 1989, at a marine ecology meeting in Sardinia,

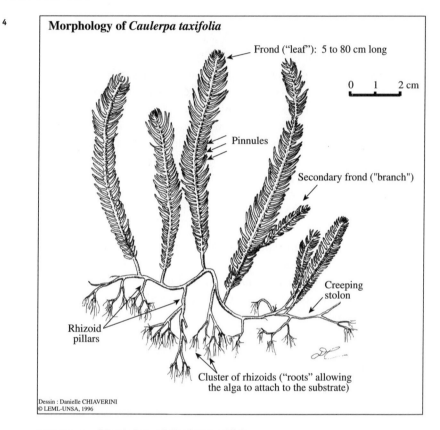

Morphology of *Caulerpa taxifolia*

Frond ("leaf"): 5 to 80 cm long

0   1   2 cm

Pinnules

Secondary frond ("branch")

Creeping stolon

Rhizoid pillars

Cluster of rhizoids ("roots" allowing the alga to attach to the substrate)

Dessin : Danielle CHIAVERINI
© LEML-UNSA, 1996

FIGURE 1.1.   Morphology of *Caulerpa taxifolia*

I first met François Doumenge, who had just been named director of the Oceanographic Museum of Monaco, replacing Cousteau. He had achieved this position after a prestigious university career. Teacher in the School of Arts and Sciences at the University of Montpellier, and deputy mayor of Montpellier, he later had undertaken, on behalf of several government agencies, many overseas missions as a technical expert or administrator. Before his arrival in Monaco, he had directed the

Vincennes Zoo and the Office of Scientific Research for Overseas Territories (ORSTOM) in New Caledonia. Trained as a geographer, he was
especially interested in my research on the impacts of coastal development and in my campaigns promoting marine reserves and the restoration of marine habitats. I therefore took advantage of our meeting to
talk to him about the introduction of *Caulerpa taxifolia.* He was already
well informed. Nothing serious, he basically told me. The alga would be
unable to spread and ought to disappear during the first harsh winter;
its origin was apparent, and we did not waste time on this point, which
seemed of no importance then.

Invited in June 1989 to lunch at his Monacan house to discuss my
methods of assessing the impact of burgeoning construction on the sea,
I told him I wanted to dive at the base of the Museum to see the full
extent of the tropical alga. He saw nothing amiss in this plan and asked
me to keep him posted on my observations. On August 28, 1989, I found
the opportunity to dive. I went to Monaco with my equipment: a "dive
scooter" (a sort of scooter for underwater areas) and my underwater
photographic apparati. The visit was both official and friendly: Doumenge knew about it, and I had the support of Commander Philippe
Roy, deputy director of the Museum, whom I knew very well.[2] He placed
at my disposal a small boat piloted by a Museum sailor, and we were
accompanied by Bézard and one of his assistants. The sea was calm and
the water temperature pleasant.

We dropped anchor under the windows of the Museum. The blocks
of cut stone, of which the immense building is constructed, are impressive, as is the way they are joined to the rocky vertical wall of the Monaco coast. Every time I see the Museum, I deeply admire its creator,
Prince Albert I of Monaco. He had financed the construction of this
monumental work while his principality, a small Mediterranean town,
was not glittering with wealth as it is today. He had constructed this
temple for science, and he made it a gift to a French foundation charged
with perpetuating the universal values of scientific knowledge.[3] He felt
such endeavors would strengthen ties among peoples of the entire
world.

6   Underwater, the seafloor ten meters down appeared blurred and all green despite the good visibility. I tried to distinguish contrasting, relieved surfaces, but everything seemed blanketed in green. I suddenly could not believe my eyes: *Caulerpa* covered everything. It was magnificent and also very surprising: I had the distinct impression of not being in the Mediterranean. Perplexed, I caressed the algae with my outstretched hand. It took me awhile to react, my emotion was so great. I photographed them and took some samples. The tremendous size of the various parts of the plant was astonishing. I moved to the right and then to the left of the large intake pipe for the seawater that feeds the Museum aquaria. With my dive scooter, I sought the limits of this magnificent prairie. We descended to twenty-eight meters; the *Caulerpa* was less dense, but it was still there.

Once I resurfaced, I asked Bézard:

"How long has there been this much of it?" Since his first observation in 1984, when he clearly remembered having found only a square meter, the alga had expanded each year.

"What was it like in the winter?" The *Caulerpa* prairie seemed to him sparser in the winter and the ends of the fronds were white.

"How far did it go? What was the area occupied?" Towards the west, the most distant colonies were found not far from the outer harbor of Fontvieille, about a kilometer away. He also had a vague recollection of a small clump of *Caulerpa taxifolia* in France, opposite Cap d'Ail, two kilometers west of the Museum in the direction of the Ligurian current. To the east, the alga had not passed the fishermen's parking lot.

The alga was basically localized in front of the Museum, where the seafloor drops off abruptly. It was growing on a surface roughly rectangular in shape, 150 meters long (parallel to the coast) and 50 to 100 meters wide (between three and thirty meters in depth): the alga now covered more than a hectare.[4]

"What do Doumenge and Roy think of this?" Since they allowed me to dive, I assumed they wanted my expert opinion.

As soon as I returned to the Museum, the very evening of the dive,

I met Commander Roy. I communicated my great concern, which could
be summarized in four points dictated by common sense:

1. The alga was recently introduced to the Mediterranean; its presence in this sea is abnormal.

2. It is adapted to all the conditions of the Mediterranean Sea and has survived several especially severe winters.[5] It will therefore not disappear because of the cold.

3. It was growing between three and thirty meters deep, on rock, sand, and mud, remarkably densely. The habitats of the alga seemed very varied. The Mediterranean species, *Caulerpa prolifera,*[6] would never have been able to colonize this zone. *Caulerpa taxifolia* is thus a species capable of competing with Mediterranean algae.

4. Its invasive nature is worrisome: from one square meter in 1984, it had grown in five years to more than a hectare. A remarkable advance!

My concerns were transmitted to the Museum directorate. Despite my alarming observations, they considered the introduction of the alga anecdotal and ephemeral and even advanced arguments in favor of its presence. According to them, the tropical *Caulerpa* was useful for the Monacan marine bottom patches degraded by the influx of urban wastewater. They reminded me that the conversion of coastal habitats during the last three decades had covered or diked more than three-fourths of the small soft-bottomed areas (between sea level and twenty meters deep) in the waters of the principality. Moreover, thousands of tons of earth had been spread in the sea in front of building sites. Thus, the Monacan coast had been reduced to a muddy, barren slope. By good fortune, these pitiful areas were now beginning to sprout a beautiful green lawn of *Caulerpa taxifolia.* In any event, the Museum directorate thought that it was already too late to eradicate it.

The serenity displayed by the Museum directorate contrasted with the growing concern of Monacan divers who saw firsthand the accelerating algal growth. Of course, I shared their opinion; it was necessary to see the invaded bottom to realize the importance of the phenomenon and the vitality of the alga. Doumenge had not dived to observe what

8   was happening under the Museum, and, oblivious to the danger, he denied it.

Faced with this potentially adversarial situation, what should I do? How should I express my responsibility as a specialist, conscious of an exceptional phenomenon that could potentially damage marine ecosystems? And how could I avoid appearing to be a "green" Cassandra? During the following months, I was often in doubt about how to proceed. After all, had not I myself advised people to plant the harmless Mediterranean *Caulerpa* (*Caulerpa prolifera*) in basins of harbors where harsh winters had caused it to disappear? But the inexplicable density of the population of *Caulerpa taxifolia* and the speed with which this alga had colonized various kinds of Mediterranean bottoms tormented me. All my thoughts tended to reinforce my conviction: this introduced species behaved abnormally.

I decided on a strategy. If the alga proliferated off the French coast, my worries would be reconfirmed. While waiting to see if this happened, I was unable to keep quiet; at least I had to suggest to relevant authorities that they undertake research on the alga, map its spread, study how it changes the ecosystems it invades, etc. All the French people with whom I discussed the invasion were surprised: the departmental Director of Maritime Affairs (Alpes-Maritimes), the bureaucrats responsible for managing the marine environment for the Department of Alpes-Maritimes, and my university colleagues at Marseilles. Some of them took to joking about it—it was just too much: a tropical alga "liberated into the sea by Cousteau," growing right under the eyes of a specialist. Others hid behind the screen of extraterritoriality, or lack of jurisdiction. It was not France, it was only in Monaco, so they could not do anything. Finally, others approved of my activities and thought I would be able to take charge of this local problem. My Monacan friends who had expertise in or authority over the marine environment of the principality were also amazed. One oversaw a society that managed the Monacan underwater reserve: the alga was not in "his" reserve, located east of the Museum. And despite his constant worry about the threatening invasion, he was unable to sensitize his superiors to the problem.

Another Monacan friend felt the potential risk was the problem of the Oceanographic Museum, which, even though in Monaco, was subject to French law. Thus, if the invasion bothered France, it was a purely French problem.

In short, everyone had reasons not to intervene, not to be involved in an affair that threatened to become highly adversarial. Ignorant of the dangers posed by this invasion, they remained cautious and uncertain. I understood their prudence but remained hopeful that some of them would decide to verify, to inquire about, to pursue the problem. I also knew that I was not yet influential enough for people to accept my evaluation of the problem. To remedy this lack of sufficient credibility, I had to bring to bear more evidence, more scientific arguments. Unfortunately, I did not have the means to undertake the necessary research on this lawn of *Caulerpa*, located in Monacan territory, in a site accessible only by boat.

All I could do was to wait for the alga to reach France. I was convinced that the observation of its spread would constitute crucial evidence of its invasive character and would awaken consciences. To watch for its arrival off the French coast, I decided to dive west of Monaco in the direction of the Ligurian current, a logical place to encounter the spread. Our university laboratory had never had the means to acquire even the smallest inflatable boat, so I had to use beaches to begin a diving exploration of small soft-bottom areas in the vicinity of Cap d'Ail. I found nothing.

In February 1990, the General Council of Alpes-Maritimes gave me a boat to use for open-sea dives. A member of the environmental bureau of the department and a student from my laboratory both accompanied me. We dove along several hundred meters of coastline at a depth corresponding to the lower limit (ca. twenty-eight meters) of the distribution of the native vegetation, meadows dominated by the seagrass *Posidonia oceanica*.[7] Beyond this depth, there is insufficient light to permit the existence of this seagrass; there is only an expanse of mud that seems sterile, so rarely are plants and animals seen. Nothing amiss, no trace of *Caulerpa*.

FIGURE 1.2.   Oceanographic Museum of Monaco

We decided to go higher in the *Posidonia* meadows as we headed back toward the coast. Still no *Caulerpa*. Although my underwater exploration had covered only a minute fraction of the soft-bottom areas located in France, immediately west of Monaco, I began to question whether the alga was really so prolific, and if it would in fact invade French waters. Beneath the Oceanographic Museum of Monaco, the tropical alga was still flourishing, slowed but not killed by the winter cold.

To the west of the Oceanographic Museum of Monaco lay a bay where *Posidonia* grew, the Bay of Fontvieille. In the early 1970s, more than twenty hectares of the sea had been filled with ballast and converted to land for construction. Nowadays the soccer stadium of Monaco is found there, as well as a new, luxurious neighborhood of several thousand dwellings, offices, buildings, and two marinas for yachts. The levee

of Fontvieille, about 1,000 meters long and surmounting soft-bottom
areas situated at depths between twenty and thirty meters, is an impressive structure. For marine organisms that live on the soft-bottom areas, it forms an artificial barrier between east and west. The species found in the small soft-bottom areas have great difficulty crossing this obstacle, unless their life cycle includes a mobile planktonic stage, as is fortunately the case for a large fraction of them. We did not know at that time if the tropical *Caulerpa* was able to reproduce sexually in the Mediterranean. If it was, would the zygotes ("eggs") be able to float westward, to surmount this barrier at Fontvieille, and to spread beyond it? Similarly, we did not know to what degree fragments of the alga would be able to detach themselves and be carried by currents to form new colonies. Whatever the reproductive mode of *Caulerpa taxifolia*, the Fontvieille levee certainly should block or at least slow its spread to the west. That could explain the absence, perhaps only temporary, of the alga in the zones I had explored.

In March 1990, as part of my university course on algae, I discussed the tropical *Caulerpa* and showed an herbarium sheet with a beautiful specimen. I asked my students and the many club divers along the coast to memorize the appearance of this alga and to inform me if they detected it in the vicinity of Monaco.

That spring, I had also interacted with a reporter from *Nice-Matin*, Christian Mars, who was always interested in the research on coastal marine life conducted in my laboratory. Transplant experiments with *Posidonia*, arguments in favor of establishing underwater reserves, artificial reefs, and the irreversible damage caused by port activities had all been covered by his newspaper. He asked if I had any new subjects that might interest him. I told him about the alga and asked him not to publicize the affair if my worst fears were not confirmed. I promised to call him if the alga reached the French coast.

Towards the end of July 1990, during a meeting of environmental organizations (NGOs), for which I had been a scientific advisor for nearly fifteen years, I ran into the secretary general of the prefecture.

12 At the closing session of the meeting, devoted to an analysis of the new burst of harbor construction projects, we discussed the alga. I still remember our exchange:

"Have you contacted the departmental director of Maritime Affairs?"

"Of course."

"What did he tell you?"

"That, for now, the alga hasn't reached France!"

He invited me to stay in contact with the local director of Maritime Affairs and to keep him posted on any changes in the situation. Once again, even though I understood this strategy of waiting and staying neutral, I regretted the absence of even the smallest initiative to launch a scientific study. An independent study would have sufficed to verify my observations and eventually to take account of the phenomenon. Isolated in my diagnosis of the situation, I should have sought concrete confirmation of the invasion. As soon as it appeared in France, I would have to make my voice heard.

July 1990: The Alga Arrives in France

All our research activities require observations or experiments *in situ*, in the field, which for us is the seafloor. In the summer, warm weather and mild water temperatures allow us to work effectively. Contrary to most of my colleagues, summer for me is not synonymous with a vacation but with intense research activities, often physically challenging ones.

During July 1990, I dove almost every day with my team as part of our experimental research on the *Posidonia* meadows. This vegetation constitutes the underwater forests that humans have greatly degraded for several years. They regenerate so slowly—a hectare destroyed at the lower depth limit of this vegetation takes three thousand years to regrow!—that we had begun propagating them by cuttings to speed their recovery.

On July 17, 1990, I found a jar awaiting me on my desk, along with this short message: "I found *Caulerpa taxifolia* at Cap Martin. Your student: Rodolphe Lemée." The contents of the jar left no doubt; this was indeed *Caulerpa taxifolia.* I was not expecting to find it to the east of the Museum, opposite the direction of the dominant current.[8] I later learned that a local fisherman set his nets at the base of the Museum and alternately at Cap Martin, five kilometers to the east. He had thus been able, inadvertently, to disseminate several cuttings. I telephoned Lemée for more information. The diving instructors of the Squale club of Menton, with which he was diving at the time of his discovery, described in detail the location of the zones covered by the alga. They then remembered having seen the alga appear at Cap Martin two or three years earlier (1987 or 1988). They kindly offered to allow me to visit the sites on the club boat.

My fears were confirmed. I had to act quickly. To get the local authorities to assume responsibility, I decided to draft a report on the matter, asking them in writing to be cognizant of the risk now threatening the marine coastal environment and to launch the necessary studies. I also decided to alert the local press. I needed the media network to inform divers quickly about the characteristics of the alga, so they could inform me in turn of its precise locations. If the alga was at Cap Martin, other capes and bays near Monaco could be invaded as well. An appeal to all those who spent time on the seafloor was necessary for further reconnaissance of the invasion.

As I set about these activities, it was Christian Mars of *Nice-Matin* who was the first to hear the news. On July 24, 1990, I informed him of the arrival of *Caulerpa taxifolia* in France. I also told a friend, Jacqueline Denis-Lempereur, then an influential reporter for *Science et Vie,* and suggested that she cover this subject in a more detailed report than the one that would appear in the local press. Several days later, I received her negative response: her editor-in-chief did not want to raise this issue.[9] On July 25, 1990, *Nice-Matin* revealed to the general public that the alga had invaded France. That very day, I was scheduled to dive at Cap Martin to describe the situation there to the reporter, but, using the information I had transmitted to him previously, he had preferred

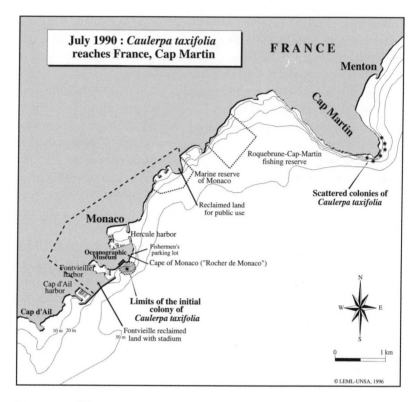

FIGURE 1.3. July 1990

to publish the "scoop" immediately. Thus, in the midst of the lazy summer of 1990, the residents of the Côte d'Azur and thousands of tourists were able to read on the front page the following gripping headline: "Ecological accident on the coast. Tentacled alga attacks coastal soft-bottom areas." And, on the inside pages: "Ecological accident on the coast. The algae attack!"

The headlines were followed, with a certain amount of exaggeration, by an exposition of the problem, illustrated by the photos I had taken a year earlier at Monaco. The Museum was directly implicated:

"Poured into the sea by the Oceanographic Museum of Monaco, the suffocating *Caulerpa taxifolia* is in the process of smothering our underwater seafloor with nothing able to stop it." The article ended: "In face of the consternation of the administration in this matter, Professor Meinesz called today for divers and bathers to be on the lookout for the advance of this alga. This is an appeal for eyewitnesses that risks embarrassing the Oceanographic Museum of Monaco. The Museum director, Professor Doumenge, in any event stated to us yesterday evening that 'on the contrary, this affair will be very beneficial for the Côte d'Azur, which will see its underwater prairies, currently in a bad way, . . . flourish again.'"

The media frenzy was under way. The divergence of opinions was not founded on purely scientific arguments. On the one hand, Doumenge is not a specialist in algae, but a respected geographer who had spent most of his career managing institutions. On the other hand, he made his statements without having seen the situation under the sea. I therefore interpreted his opinion as a simple defensive reaction to an affair that was now highly publicized. Doumenge had apparently sought to minimize the gravity of the situation, without denying the origin of the introduction. To my knowledge, he had not changed his opinion since our last meeting.

The headlines and certain exaggerated passages in the article shocked me more than Doumenge's response. Mars is known for taking confrontational positions, and it is always very hard to ask a reporter to write a correction to an article. But some paragraphs were so excessive that I telephoned him the same day to suggest a clarification. I asked him why he had written that "the affair troubled scientific circles" when he knew full well that, except for me, no one was worried about it. I cited phrasing in the article that went well beyond my statements: "the alga swallowed whole our great blue sea . . . the invader is everywhere . . . this plague . . . this plant inexorably smothers the environment with its tentacles." He was above all mistaken on the mode of action of the toxicity of this alga, declaring "its irritating toxins make fish and crustaceans flee."

16   That afternoon, I telephoned Doumenge to tell him of my reservations about the article and to reassure him about my intent. I told him about my interactions with Mars. He informed me that he had just faxed him a correction and confided to me that the article had really embarrassed him because he was indirectly implicated, even though he had nothing to do with the affair. In fact, the alga had been "dumped" into the sea during the period when Commander Cousteau, his predecessor, was director of the Museum.

The next day, in the same newspaper, appeared a clarification, again signed by Mars. The headline still emphasized the difference of opinion among scientists: "Scientific controversy over the tentacled alga. The Oceanographic Museum of Monaco defends the beneficial nature of the invasion." The first part of the title displeased me: barely exposed, the risk to the environment seemed already contested and thus belittled. Caught overinterpreting statements, the reporter hid behind a scientific pseudo-controversy in order not to seem to contradict his previous report. Lacking my comments, the clarification resembled a unilateral rebuttal in favor of the Oceanographic Museum.

The article recalled the origin, still uncontested, of the invasion: "This tropical alga that, since its escape from one of the tanks of the Oceanographic Museum of Monaco, is proliferating on our banks." This view of the origin was confirmed by the clarification of the Museum director: "In fact, these algae have long carpeted many tanks of the Museum aquarium without harming their flora and fauna. In this regard, it should be noted that, for many years, thanks to the implantation of *Caulerpa* at the foot of the maritime façade of the Museum, species like *Octopus macropus,* moray eels, wrasses, rainbow wrasses, and damselfish have returned in great numbers to inhabit the zone that they had formerly abandoned. These animals," argued Professor Doumenge, "have, thanks to the colonization by *Caulerpa,* which emits no irritating toxin, found veritable prairies where there had previously been nothing but desert. Unfortunately, this implantation will doubtless be reduced to nil during the next winter that is a little colder than the preceding ones."

It was evident that Doumenge was minimizing the problem and

hoping it was temporary. The title of the article summarized his opinion 17
well: "The Oceanographic Museum of Monaco defends the harmless
nature of the invasion." But I was troubled by his reasoning. I knew that
his assertions rested on no specific research. To my knowledge, not a
single census of fishes had been taken by Museum divers before or after
the establishment of *Caulerpa.* In predicting the imminent demise of
the alga, he dodged the essential point: the risk of its spread, to the
detriment of coastal Mediterranean ecosystems. On the other hand, although species in the genus *Caulerpa* were known to contain toxins,
no one had ever attributed to them the property of "emitting irritating
toxins," an unfortunate characterization for which I had reproached the
reporter. On this issue, the "controversy" was over an error introduced
by the latter.

In this second declaration to the local newspaper, one detail should
be emphasized: at no point did the director of the Museum deny the
origin of the alga. He confirmed that it had existed for a long time in
the Museum aquaria and even bragged about the merits of this "implantation" and explained how the first colony had been followed carefully by Museum staff. "Implantation!" Doumenge used the word in
his statement faxed to *Nice-Matin.* Implantation was to be evoked two
years later.

Only two local newspapers, *Var-Matin* and *Marin*, published extracts from the articles in *Nice-Matin.* The story remained largely local
and anecdotal, and its impact had been lessened by Doumenge's contradictory suggestions. Jean-Pierre Lamouroux,[10] of the regional television
station FR3–Côte d'Azur, presented a short report on the subject during
the regional news. He had labored to film in person the rare, sparse
clumps of *Caulerpa taxifolia* at Cap Martin. His underwater shots today
serve as benchmarks on the state of this region before the general invasion of the site; Cap Martin itself turned out to be barely touched.

I hastened to finish the report in which I summarized my knowledge on the subject. On July 31, 1990, I transmitted it to the prefect
of Alpes-Maritimes and the president of the general council of this department, with a copy to the departmental director of Maritime Affairs.

18   Henceforth, those in charge of managing our environment had in their hands all the facts then known about the alga *Caulerpa taxifolia.*

The conclusions of the report were as follows:

1. It is certain that the alga introduced six years ago at Monaco is well adapted to the most species-rich habitats of the Mediterranean at depths of three to thirty meters. It weathered the severe winters of 1985 and 1986. The alga is beginning to grow rapidly on both sides of the original introduction. Because of its rapid growth and great reproductive potential, one can expect a broad and rapid increase in its range.

2. If its advance is at the same rate as that observed until now, there is a risk that this species will replace many native species: hundreds of different species live between three and thirty meters deep, and a great number of them play key roles in the food chain.

3. We do not yet know which native animal species are able to consume the alga. Our preliminary observations show that several fronds are lightly grazed. Neither do we know the possible concentrations of the toxins (caulerpine and caulerpicine) that this species of *Caulerpa* contains.

We believed the following studies should be urgently undertaken: 1. Monitoring of the advance of the alga and a study of the characteristics of its reproduction and speed of growth. 2. A study of competition between this *Caulerpa* and the algal species in the Mediterranean. 3. Behavioral studies on the grazing of this alga by native animals. 4. Biochemical studies on the concentrations of the toxins.

My dispatch[11] elicited no response, not even so much as an acknowledgment that it had been received. I also sent the memorandum to Doumenge, whom I hoped to win over, and I strove to be conciliatory in my cover letter. After all, it was true that tropical fish live in aquaria with *Caulerpa* species and feed on them sometimes; it is also true that the alga is beautiful and the soft-bottom areas under the Museum that are colonized by the alga are magnificent, modified though they are— into underwater golf courses. My letter, as with the previous ones, received no reply.

September 1990: The Alga is Found at Toulon

The several articles that appeared in the local press and the televised report at least had the beneficial effect of making the alga a celebrity. In the following months, I harvested the fruits of this reportage. I received several notices of new *Caulerpa* colonies; thanks to the news reports, I was able to advance my understanding of its spread.

In September 1990, I received a letter from Toulon, accompanied by a dried fragment of *Caulerpa taxifolia* and an excellent diving report. On the regional televised news, my correspondent had learned of the appearance of the alga at Cap Martin. The day before, diving in the Méjean cove near Toulon, 150 kilometers west of Monaco, he had noticed an unusual alga; it resembled the one shown on television. He returned to the spot and collected a fragment, which he sent to me with a precise description of the occupied area (several square meters) and a map of its location in the bay, an anchoring site highly favored by recreational boaters. We arranged a meeting to dive. At the Méjean cove, on September 7, 1990, I met Pierre Granaud, a big fellow, a retired photographer and amateur naturalist. The appearance of the beautiful alga at the soft-bottom sites he regularly visited had not escaped his notice. His naturalist's sense had inspired him to dive again to confirm and detail his observation. With a friend, Jean-Marie Astier, a correspondent, diver, and naturalist from Toulon, it took us more than half an hour to find the main colony of four square meters and to observe that a dozen small satellite colonies were spread around it, within a radius of ten meters.

How had the alga reached this cove? Transported on a boat anchor? By the floating stages of its life cycle? If the latter were true, it ought to have been everywhere. This first appearance of *Caulerpa taxifolia* on the shores of the Department of Var was worrisome. Given the immensity of the underwater area to be explored, it was absolutely necessary to sensitize the thousands of scuba divers and snorkelers who visited these coasts each year, so that they would report all sightings.

Jean-Marie Astier was well acquainted with Gabriel Jauffret, reporter for *Var-Matin*. We had guessed that it would be useful to inform

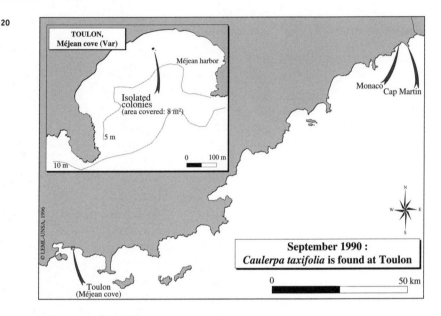

FIGURE 1.4.　September 1990

him of this discovery. It was he who had signed the article on *Caulerpa taxifolia* that appeared in the *Var-Matin* of July 28, 1990; he had gone over the information given by *Nice-Matin* and had interrogated IFREMER,[12] which had, as I had, disavowed the exaggerations regarding algal toxicity. However, the spokesperson for that institution, chief of the aquaculture laboratory at Nantes, had raised a surprising argument: the alga is harmless, because it is edible! Recalling this fact to Jauffret, I invited him to eat a fragment of the alga. I had already performed the experiment: at first, the alga is bland and flavorless, but, after a minute, a disagreeable sensation seizes the palate, tongue, and back of the throat. Smarting, a peppery taste, slight numbness of the tongue: all the tasters have the same experience and none likes it. They quickly spit out whatever remains and reach for a large glass of water.

Other species of the order Caulerpales are also inedible. In the east-

ern and southern regions of the Mediterranean where there is much
Caulerpa prolifera (the common *Caulerpa* in the Mediterranean), a fish
known as salema (*Sarpa salpa*) can cause hallucinations in those who
eat it at certain times of the year. Several authors agree that the cause
of this problem is the diet of salema, which consume the alga in these
regions during the autumn. In tropical seas, similar effects augmented
by loss of consciousness or reversible local paralysis have also been attributed to the direct ingestion of certain tropical species of Caulerpales.[13]

Doubtless convinced by these data and above all by the unpleasant
taste of what he had just chewed on, Jauffret headlined his article of
September 8, 1990, "Discovery of a toxic alga in Méjean cove at Toulon."
On an inside page, one could read: "A hallucinogenic alga. A tropical
alga accidentally introduced to Monaco in 1984, carried to Méjean cove
by the Ligurian current, threatens to replace local species and contains
two toxins that can accumulate in the flesh of fishes."

Although the risks were described in the conditional tense, the news
must have scared more than one fisherman from Var. On September 14,
1990, a short paragraph in the daily *Var-Matin* recalled and confirmed
the presence of the alga at Méjean. While emphasizing my suggestions
(described precisely in my report), the reporter took up the opinions of
Doumenge and the spokesperson for IFREMER: "The advance of the
alga should be noted, and studies should be conducted on the grazing of
this alga by the local fauna and on the concentrations of the toxins it
contains. The director of the Oceanographic Museum of Monaco feels
that it is not dangerous to humans and the aquaculture laboratory of
IFREMER states that it is eaten in the form of a salad or as a condiment
in certain countries of the South Pacific."

After this clarification, the reporter renewed the appeal to divers to
inform me of possible occurrences of the alga along the Var coast. This
calming restatement of the problem was necessary, and Doumenge was
correct to say that the toxic alga is not dangerous to humans, given the
current state of our knowledge and the small occupied area. On the
other hand, I found the statement by IFREMER to be out of line and

22   misleading; they had minimized the problem once again, affirming that the alga was edible, therefore harmless.[14] I was burned by the positions my opponents had staked out, which were expressed by media middlemen and by people who could hardly have been said to have studied the problem in the field. The exaggerations of reporters had distorted my remarks and thus disrupted a calm evaluation of the problem. The whole affair was becoming more and more burdensome, but I felt a responsibility to communicate what I had learned and my understanding of the situation. Having fought for twenty years to preserve coastal ecosystems of the Mediterranean, I could not just let this alga invade the entire region without reacting. Mainly, I was disappointed by the indifference of the authorities charged with protecting our environment. Despite my demands, no surveillance program was established, no informed alternative plan adopted.

In order not to appear a media-mad preacher and mystic, I had to suspend all contact with reporters for a while and learn to convince the authorities in a different way. No competent scientist had invalidated my evaluation, but I was forced to seek further evidence. Though it would mean letting several precious months pass during which the alga would disappear from the public's radar screen, I knew I had to produce a scientific publication on the matter.

February 1991: The First Scientific Publication

From the beginning of this affair, I had begun to amass all the elements of a scientific paper. This demanded time, fieldwork, and a complete analysis of the scientific literature.

From September 1990 to January 1991, we had increased the number of dives at Cap Martin. We had to measure the advance of the alga, gather vegetation samples in marked areas in order to calculate the density of various vegetational components, collect herbarium specimens, describe morphological variation, and photograph the alga in its new habitat. The cold, the limited range permitted by the capacity of our air tanks, and the easterly wind often interrupted our work, forcing us to repeat dives. Back in the laboratory, we consulted hundreds of bibliographic references on the Caulerpales.

There was no specific financial assistance for this research. My laboratory, consisting of four teaching researchers, was at that time supported by none of the major research institutions. We had to bootleg the necessary funds to finance this basic work from applied research contracts: studies of impacts of coastal conversion, or of transplanting *Posidonia*. It was with these funds that I was able to undertake the first basic research on *Caulerpa taxifolia*.

At the beginning of November 1990, divers of the Environmental Directorate of the Department of Alpes-Maritimes alerted me to the presence of the alga in the fishing reserve of Roquebrune–Cap Martin, located between Cap Martin and Monaco. They invited me to dive to verify the finding. In fact, five meters deep, three dense colonies of *Caulerpa* from one to two meters in diameter were growing on rocks. Coincidentally, six months earlier, I had mapped this zone at the behest of a local association for environmental protection.[15] During that detailed exploration of the site, we had not found the alga. I was therefore sure it had arrived there in the interim and had grown to its current size in less than six months. Despite this new discovery, I decided not to go public with my arguments; I refused all new contact with both the authorities and the news media. I did not want to find myself battling a new contradictory statement from the director of the Oceanographic Museum of Monaco.

For the dives, I had the technical support of Bruno Hesse, a conscientious objector assigned to our laboratories by the Ministry of the Environment. He was an excellent diver who, in a former life, had occupied the unusual role of killer-whale trainer at Marineland in Antibes. He accompanied me on all my underwater excursions at Cap Martin. To get to the water, each time we had to carry forty kilograms of equipment on our backs along the footpath at Cap Martin. We were often bruised and scratched, and knocked about by heavy swells amidst the sharp rocks. My partner stayed under water for hours to ensure my safety while watching me photograph or manipulate *Caulerpa*. It was in memory of these physically taxing dives and to thank him for his enthusiastic help that I made him coauthor of the first scientific publication on the introduction of the tropical *Caulerpa* to the Mediterranean.

Oceanologica Acta published the article, "Introduction and invasion of the tropical alga *Caulerpa taxifolia* in the northwestern Mediterranean," transmitted in February 1991. The journal is edited by IFREMER, and I hoped by this publication venue to attract the attention of the upper echelon administrators of this institute, which is the obligatory dispenser of expertise to the French government on anything having to do with the seas. Before submitting the manuscript, I asked the director of publication about the publishing lag time. He guaranteed a rapid review and, if the opinions of the referees were favorable, publication within eight months, thus before the autumn of 1991. For nonexperts, such delays can seem very long. They are primarily caused by the slowness of the referees, because, as for any respected scientific journal, each manuscript received is subjected to analysis by two experts who are chosen for their competence in the subject and volunteer for the task. In general, these referees, who remain anonymous to the author, do not spare the author's feelings; scientists often endure categorical rejection or are forced to revise heavily and to correct certain points. Our article received three reviews: the subject matter was so surprising that the editor wanted his verdict supported by an extra referee. The three reviews were favorable.

For two years, I had been describing to local authorities the many ecological risks I had found associated with the invasion. To make these new elements known quickly, I asked the editor-in-chief of the journal for permission to transmit the in-press paper to the relevant ministries. The following pages and the appendix summarize the key information contained in the publication,[16] to show what the French authorities knew before the spring of 1991.

One of the points discussed in this article concerned the introduction of algae into new ecosystems, and the means by which these introductions occurred. Fewer than fifty species of algae are known to have

been introduced into the Mediterranean (of a rich Mediterranean algal
flora of more than a thousand species). The great majority of these nowadays occupy very restricted ecological niches and do not compete with
native species; they therefore do not damage ecosystems. But there is no
comparison between the behavior of a green alga like *Caulerpa taxifolia*
and that of a red or brown alga. The origins of these introduced species
need to be identified because, even if such species cannot be controlled,
at least their appearance could then be understood.

Species have followed three primary routes to the Mediterranean:

1. *The continual introduction of species from the Red Sea since the
opening of the Suez Canal more than a century ago.* The majority of these
introduced species (of which forty are algae) remain restricted to the
eastern Mediterranean and, more narrowly, the coasts of the countries
located to the east of the canal, in the direction of the dominant currents. These introductions are a direct consequence of a human act: the
construction of the canal.[17] The abrupt acceleration of the pace of introduction observed for the past few decades in the Mediterranean is the
result of another human construction: the Aswan Dam. By appreciably
reducing the flow of water from the Nile into the Mediterranean, humans have modified the salt content of the Nile delta waters, right at
the mouth of the Suez Canal. Before the construction of the dam, the
mixture of the fresh water from the Nile with seawater did not allow
the majority of marine species coming from the Red Sea to survive. By
the same token, today there is a water deficit in the eastern basin of the
Mediterranean: in the summer, evaporation surpasses the arrival of water from rivers and from the Atlantic Ocean, via the Straits of Gibraltar.
The slight depression formed in the eastern Mediterranean produces a
current in the Suez Canal from the Red Sea towards the Mediterranean.
This is how the marine flora and fauna of the eastern basin of the Mediterranean are being modified, but, until now, without profoundly alter-
ing the proportions of species of various biogeographic origins. The advance of tropical algal species thus introduced is slow: in 1990, none had
yet penetrated the western basin of the Mediterranean. None has proven
to damage populations of other species, each having found a narrow,

26   previously unoccupied ecological niche; they have thus enhanced local biodiversity.

2. *The inadvertent introduction of species by international naviga-tion.* Algae contained in ballast water or attached to the hulls of ships can also travel and "contaminate" all seas. Several species, today very common in the Mediterranean, were probably introduced more than fifty years ago by this means. Among species appearing more recently is the small red Australian alga, *Acrothamnium preissii*, which appears to have been introduced in this way. It was found for the first time in 1969 on the Italian coast in the vicinity of Leghorn (Livorno). My phy-cological[18] colleagues from the CNRS and I discovered it in 1982 in France, at Villefranche-sur-Mer, near Nice. Other colleagues noted its presence in 1984 still further west, at Saint-Tropez. This example of in-troduction and spread well illustrates how even a small alga, difficult for a non-specialist to distinguish, is spotted as it moves from place to place, thanks to the efficiency of the network of Mediterranean phycolo-gists.

3. *The recent introduction (since 1969) of algae attached to oyster spat imported from Japan.* Among these species, three are large brown algae that can surpass two meters in length. Introduced to the ponds of Lan-guedoc, they are currently beginning their advance in the open sea to-wards Spain. The most prolific is a big sargasso *(Sargassum muticum)* that also invades the French Atlantic coast and the North Sea. The two others, *Laminaria japonica* and *Undaria pinatifida*, are edible and are even cultivated in Southeast Asia. They were recently imported to Brit-tany for cultivation, which has led to the initial invasion of these coasts. The proliferation of these three algae is spectacular only during certain seasons and with certain configurations of particular soft-bottom areas. In France, their implantation has already been deplored by cultivators of oysters and mussels in the ponds of Languedoc, because these algae attach to the platforms set out for the shellfish and hinder their storage and manipulation.

When the imported algae proliferate and compete with the indige-nous flora (as is the case for the large Japanese sargasso in northern European seas), scientists term them "aggressive invaders," and the invasion is called "biological pollution." The international scientific community has often wished to eradicate such species. This has been attempted for the large Japanese sargasso in several British bays. But efforts at extirpation have been in vain, started much too late.

What about *Caulerpa taxifolia?* To describe the characteristics of its adaptation and proliferation, it is necessary to mention the origin of its introduction. This was a touchy question that we had deliberately broached in prudent terms in our first scientific article. As of February 1991, the means of introduction of *Caulerpa taxifolia* to the Mediterranean was not questioned; even Doumenge had not contradicted the fact that the alga had "escaped" from the Oceanographic Museum. He had even emphasized in *Nice-Matin* that the alga had long been grown in the Museum aquaria and that for several years Museum personnel had observed fishes use the colony implanted in front of the Museum.

The testimony of the assistant director of the aquarium, the presence of the alga in the aquaria several meters from the sea, and the growth of the small colony observed in 1984, noted by Museum divers, constituted a body of coherent evidence of the introduction of *Caulerpa taxifolia*, previously unknown in the Mediterranean. When I sent my manuscript to the editor, the origin of the alga was not a hypothesis but a fact, admitted even by the Museum directorate. If one finds a rare specimen of an Australian eucalypt in Auvergne just outside the wall of a botanical garden where the species thrives, one hardly has to ask how it got there.

Discreet and conciliatory, we nevertheless mentioned its origin (in the conditional tense) in the publication.[19] It was logical and useful to mention it, if only because this fact constitutes an important scientific element to help understand the features of the dissemination of the alga from its highly restricted point of introduction. It also emphasizes the risk posed by a new mode of introduction of exotic species: that of aquaria. At this time, public aquaria, installed on international coasts and functioning with pumping and discharge of seawater, were multiplying.

28   In the article, we also detailed the available information on the anatomy and morphology of this beautiful stranger. By the same token, we described its behavior in the Mediterranean: its reproduction, its ecological niche, and its tolerance of winter temperatures. Finally, we presented several lines of evidence on its identification and on the other *Caulerpa* species in the Mediterranean (extensive extracts of this publication are found in the Appendix). The last section of the publication was devoted to a sensitive subject: the possible toxicity of the alga. Although many terrestrial plant species are toxic, among large algae just a few relatively rare families are known to accumulate toxic substances that play a defensive role against herbivores. This is the case in the Caulerpales. To present this subject, we stuck to salient points from the scientific literature. The data justified concern about potential effects of what would probably turn out to be a degree of toxicity of *Caulerpa taxifolia*. Thus, two traits of this alga underlay our concern: the dominance of this invasive species over the native Mediterranean flora and its likely toxicity. We were not confronted simply with the anecdotal arrival of one more toxic plant in a habitat dominated by nontoxic species, but with the invasion of a toxic plant that constituted underwater "forests" and that allowed only rare nontoxic plants to survive. In this scenario, the exact degree of toxicity is not of great importance, but this point nevertheless became a subject of great controversy in the popular media.

The conclusion of our publication was that *Caulerpa taxifolia* presented all the characteristics of a noxious alga and threatened coastal Mediterranean ecosystems. Our text was buttressed by eighty-one scientific citations. After submitting the manuscript in February 1991, I felt I had accomplished more than a routine study, more than a small increment of knowledge. I had described a risk for Mediterranean ecosystems. As one of the few specialists on the subject, I had the moral responsibility to publicize this risk. At the time, I was persuaded that the publication of this study would end the useless polemics and would induce the authorities to deal with the problem effectively. It was necessary to act quickly—the alga already covered three hectares of seafloor.

The Alga Grows...and the Polemic Begins

If 1990 was the year of observation and reports, 1991 was essentially devoted to the authorities. My determination to convince the authorities of the crisis grew progressively stronger. Above all, it was continuous self-criticism of my observations and deductions that determined my actions. Was I right or wrong? I had skimped on my efforts to deepen knowledge of the subject and to search for collaborators outside my laboratory. My adversaries' adoption of dogmatic and intolerant positions barely touched me; I viewed them as just so many manifestations of scorn for the environment. They catalyzed my determination to persevere. Of Dutch origin, one of my faults is that I am stubborn. My Dutch ancestry prompted me to apply the motto of the rulers of the House of Orange: "I will stay the course!" My aim in this affair was to make known a situation that had to date been belittled and neglected.

Foreseeing a difficult debate, I had compiled, starting in early 1991, all writings concerning *Caulerpa taxifolia*. This task was essential to grasp the meaning of this delicate period. Certain events or data, appearing minor or even unconnected to the subject at hand, were nevertheless part of the story of this invading alga. It is thus with all due precision that a number of facts will be recalled here, facts that are not

30   widely known but that are useful in understanding the behavior and
decisions of the actors in this affair, including the alga itself.

February through April 1991: The Authorities are Alerted

My appeals in July 1990 to the prefecture, the General Council, and
Maritime Affairs of Alpes-Maritimes received no response and had no
impact. In February 1991, bolstered by the scientific study supporting
my initial analysis, I felt it necessary to alert the authorities once again.
This time, though, I adopted another strategy and consulted local contacts in the two most relevant ministries.

At Aix-en-Provence, I met the specialist on coastal problems for the
Regional Directorate for Architecture and the Environment,[1] an agency
representing the Ministry of the Environment in the region encompassing Provence, the Alps, and the Côte d'Azur. Convinced by my arguments but powerless to make decisions, he advised me to assemble a file
and to present it to the central services of his ministry. However, he felt
this surprising tale, presented by an isolated academic, had little chance
of convincing people. Given the various elements of the case, he suggested that it would be difficult to get the upper-level bureaucrats of the
Ministry of the Environment to bring expertise to bear on the problem.
It would be better to submit a request for funding to generate a deeper
understanding of the subject, as that would force some departments of
the ministry to analyze the problem. I thus had to create a research
project to fill gaps in our knowledge, a project that would have to be
submitted in association with colleagues of other laboratories. He advised me to submit the proposal to the director of the research department of the ministry.[2]

The message was clear; I was too isolated and too impassioned to
engage the interest of the people I talked to. To have a successful proposal, I first had to convince other scientists to collaborate in planning
a joint research project. This is how I came to approach Charles-François
Boudouresque, director of the laboratory of marine biology and benthic

ecology in the faculty of sciences at Marseilles-Luminy.[3] We had collaborated for almost fifteen years on research programs on ecosystems in
the marine reserves of Corsica and in the Port-Cros National Park. The
dozens of joint research efforts, the memories of diving together in extraordinary places, and our passion for the protection of coastal ecosystems of the Mediterranean had woven strong ties of friendship and mutual esteem. He agreed to add to our project a person we considered
one of the leading phycologists of the Mediterranean, Marc Verlaque,
research associate at the CNRS and affiliated with Professor Boudouresque's laboratory. Verlaque accepted enthusiastically. He developed
a part of the proposal to study competition between algal species, particularly *Caulerpa taxifolia* and the other algae in the Mediterranean.

To study the toxicological aspects of the problem, I turned to the
department of the National Institute for Health and Medical Research
(INSERM) that is charged with studying the pharmacological or toxic
properties of products of the sea, based in Villefranche-sur-Mer, near
Nice. I knew a researcher there, Danielle Pesando, who agreed to draw
up a preliminary program of studies on the mode of action of the *Caulerpa taxifolia* toxins.

I was assured of the collaboration of excellent specialists of national
repute that would allow us to study in depth the two themes most crucial to the determination of the potential risks associated with the algal
invasion. For my part, I drafted a program of studies on the biology of
the alga in order to understand its invasion strategy better.

In the meantime, I consulted with the director of Maritime Affairs
for Alpes-Maritimes, the local representative of the Secretariat of State
for the Sea. He was impressed by the strategy chosen to sensitize the
Ministry of the Environment to the problem, and he advised me to present the same project in Paris to an assistant of the Secretariat of State
for the Sea. Increasingly convinced that something abnormal was happening—fishermen had told him about hauling up *Caulerpa* from
depths of greater than a hundred meters[4]—the director of Maritime
Affairs took the initiative, at the beginning of April 1991, of seeking the
opinion of IFREMER experts based in La Seyne-sur-Mer, near Toulon.

To justify his request, he sent them, with my authorization, a copy of the study in press in *Oceanologica Acta.*

It was easy enough for me to obtain appointments with people I sought out in the two ministries. On April 14, 1991, I flew to Paris with the *Caulerpa taxifolia* file in my suitcase. In the morning, I met the director of the research department of the Ministry of the Environment. To my great surprise, he began speaking to me of an old letter about an entirely different matter.[5] After telling him the goal of my current visit to Paris, I had to explain to him in a quarter of an hour the threats the alga posed to the coastal environment. In closing, I presented him with the research proposal for studies to be undertaken with great urgency.[6] Disconcerted by my story, he called in one of his colleagues to whom I had to repeat my exposition while my original interlocutor occupied himself on the telephone with another matter. The conclusion of the interview was that my research proposal was seen as interesting, but funds were not available in a short period to finance it, and he suggested contacting the Secretary of State for the Sea.

The weather was good in Paris. Leaving the Ministry, I enjoyed contemplating the Seine. I was optimistic, but I was also very wrong.

In the afternoon, I met with the representative of the Secretary of State for the Sea, who at that time was Jean-Yves Le Drian. After listening to me for five minutes, he left, citing an extremely urgent report he had to prepare for the Secretary. He passed me over to the head of research for the Interministerial Committee for the Sea. My visit was not scheduled in the very crowded docket of the latter bureaucrat. After yet another exposition of the problem, the same scenario was reenacted: someone telephoned him, he left the office, returned, and left again as he made his excuses, all the while introducing me to his collaborator, a woman newly assigned to this office. I was not stupid. This ritual of passing the torch to interlocutors of increasingly lower status annoyed me. I had come from Nice to explain to a high-level bureaucrat a potentially very worrisome risk for the Mediterranean, and my reception was disappointing, to say the least.

The bureaucrat of the Interministerial Committee for the Sea understood my disappointment. She listened to me with interest and admired the herbarium sheets I had brought along. Without prejudging the final opinion of her superiors in the bureaucratic hierarchy, she explained to me, exactly as had the director of the research department of the Ministry for the Environment, the low likelihood that her agency could intervene and fund such research. Walking outside to the Place Fontenoy, I nevertheless remained optimistic. I was certainly overconfident.

The problem posed by the alga, a product of the marine public domain, fell under the exclusive jurisdiction of the national government. I had indeed contacted all the authorities who might have been able to take charge. The local communities were also able to intervene, either by transmitting demands and pressure to the national government, or by financing operations concerning the development or protection of the environment. My first letter, addressed to the President of the General Council of Alpes-Maritimes in July 1990, remained unanswered. However, I asked nothing more than an expert evaluation of the situation.

The most urgent task remaining was to gather all the data on the spread of the alga. I would have to increase the number of dives on both sides of Monaco and sensitize the divers of the Côte d'Azur, and I therefore asked the General Council of Alpes-Maritimes and the Regional Council of Provence–Alpes–Côte d'Azur for financial assistance. With unusual speed, my proposal was accepted. Thanks to this support,[7] I was able to produce and to publish seven thousand copies of a brochure that briefly introduced the alga. *Caulerpa taxifolia* was described with color photographs to allow identification. All divers were asked to notify us of colonies observed in the course of their underwater excursions. We then recorded all discoveries in a central database and validated each one by diving at the site.

These brochures were distributed to all the diving clubs on the coast, and I began to assemble a network of volunteer informants. The publicity campaign was launched in the local press on April 11, 1991, with a very calm and even reassuring article in *Nice-Matin*. Finally the

34   problem was presented to the public without exaggeration or controversy. Invited to mobilize themselves to aid us in tracking the tropical alga, the divers of the Côte d'Azur reported several new colonies, more or less distant from the *Caulerpa taxifolia* prairies known in 1990. Thanks to local organizations, to the media personnel, and to the civic-mindedness of many observers of the marine environment, we were finally able to map the spread of the alga.

April 1991: The Director of the Oceanographic Museum of Monaco Gets Angry

At the end of 1988, Doumenge had taken the helm of the Oceanographic Museum of Monaco from Commander Cousteau. He had to take careful control because the number of visitors to the Museum had begun to decline. The newspaper *Nice-Matin* reported on the new strategy, "Oceanographic Museum of Monaco. Second wind. Professor Doumenge intends to 'dust off' the venerable institution."[8] To regenerate interest in visits to the aquaria, the new directorate chose to improve the exhibits of tropical species and, especially, those of corals. It also planned to diversify the extension activities of the Museum by internationally marketing assistance in setting up public aquaria and expertise in culturing corals.

    This new orientation was to flourish thanks to the help of the biologist Jean Jaubert, specialist in tropical corals and aquarium enthusiast. As assistant professor at the University of Nice–Sophia Antipolis, he had begun to elaborate this research agenda. He first applied his know-how in the emirates and monarchies of the Middle East, thanks to a long succession of assignments at the Ministry of Foreign Affairs. After a rapid return to his duties as teacher-researcher at the University of Nice, he was promoted to professor in 1988. From this date, he based his activities at the Oceanographic Museum of Monaco. A large part of his research was aimed at development of a very clever aquarium procedure, patiently brought to fruition at the Faculty of Sciences of the University of Nice. This procedure, called "Micro-ocean," allowed the survival and growth of tropical corals in an aquarium set up as a closed system. He patented the technique as a private citizen in June 1990.[9]

During the summer of 1989, several tons of corals[10] were removed from the borders of the Indian Ocean and the Red Sea, packed at Djibouti on the Somalian coast, and shipped at great expense to Monaco. The surviving corals constituted the marvel of the Oceanographic Museum of Monaco in an aquarium of forty cubic meters constructed especially according to Jaubert's procedure. The local and national press acclaimed this attraction.[11]

In 1990, at the urging of Doumenge and Jaubert, the Principality of Monaco created an entity managed by the Monacan Scientific Center: the European Oceanographic Observatory.[12] Located at the Oceanographic Museum of Monaco, it consisted at the outset of two researchers and three graduate students directed by Jaubert, who had obtained a new assignment in higher education. The first research conducted at the Observatory was based on experiments on corals carried out in the closed-system aquaria. According to Doumenge and Jaubert, research on coral physiology and growing techniques would allow their cultivation, with the goal of compensating for the growing restrictions on the importation of living corals for the aquarium trade. The distribution of the Micro-ocean procedure, especially effective for corals, required this first inoculum, still too rare in aquarium shops.

The large tropical aquarium of the Museum and the research aquaria of the Oceanographic Observatory became an excellent means of promoting Micro-ocean. The aquarium techniques were offered to various countries of the Arabian Peninsula,[13] the Americas, and Southeast Asia.[14] This initiative, logical for an institution that had to fund its operation, allowed for a diversification of economic resources. *Science et Vie* characterized the operation: "As was emphasized by Professor Doumenge, director of the Oceanographic Museum of Monaco and the International Commission for the Scientific Exploration of the Mediterranean, mastery of reproduction by corals and their associated species would feed the large aquarium market, which would prevent the pillage of tropical reefs. Micro-ocean is therefore profitable. . . . The Japanese and the Americans are already buyers."[15]

In February 1991, when the Gulf War ended, hydrocarbon pollution seemed to threaten the ecosystems of the Persian Gulf. It was easy to observe that the petroleum poured into the Gulf in some places mucked up small coastal soft-bottom patches. The gigantic smoke clouds emitted by the burning oil wells could also perturb local environmental conditions (by creating less light and lower water temperature) crucial for the lives of tropical species. The risk of a major blow to marine biodiversity was much bruited in the French popular media. The situation, described by several scientists, seemed catastrophic.[16]

Jaubert, honoring the accords with the Council of Europe, which had enjoined his observatory to work for the prevention of major risks to the marine environment, traveled to Kuwait in late winter 1991. He painted a picture of the Gulf ecosystems for the press, emphasizing the threat of global extinction of a large number of coral species found only in the Persian Gulf.[17] For his part, Doumenge estimated that, in Kuwait, humans were threatening animal species—corals—that he deemed more vital to the planet than tropical forests. In press statements, he linked the threat of coral extinctions in the Gulf to the global problem of CO_2 increase: "Barrier reefs oppose the greenhouse effect that afflicts our planet because of an excess of carbon dioxide."[18] "Tropical forests absorb only 25 percent of the atmospheric CO_2; the rest is fixed by corals."[19]

These alarmist statements in the press, relayed by environmental non-governmental organizations with an antimilitary bent, led the United Nations and the countries engaged in the Gulf to take measures to limit environmental degradation. Environmental impact assessment and restoration of coastal marine ecosystems constituted a lucrative market for oceanographic teams from all over the world. Many offers of expertise arrived in Kuwait, Saudi Arabia, and the offices of the United Nations.

In early February 1991, Jaubert advised the Saudis to "place several coral specimens in a swimming pool" with a view to "reestablish the coral reefs after the catastrophe."[20] From this idea, a project to safeguard the corals of the Gulf was concocted. It comprised, among other facets, a campaign to harvest species of corals "threatened" with extinction, to be cultured in tanks constructed especially for this purpose in the sea at Oman, on the other side of the Arabian Peninsula in the Red Sea, and even further away, including Italy and the Oceanographic Museum of Monaco. After having been saved and pampered in these aquaria, the corals would be replanted in the Persian Gulf. Two years of husbandry were envisioned. This would allow time for the reduction of pollutants in Kuwait and the cleansing of the habitat. The bill for this operation—termed "Noah's Ark"—submitted in March 1991 by the European Oceanographic Observatory of Monaco to international organizations[21] rose to more than five million dollars, of which four million were entirely devoted to the construction of giant aquaria modeled after the Micro-ocean principle!

The rescue operation was directed by Jaubert's European Oceanographic Observatory, funded at that time by the Council of Europe, but the Observatory technically reported to the Monacan Scientific Center and therefore to the Principality of Monaco. These particular corals, scleractinians, had just been reclassified from list 1 (very restricted export) to list 2 of the convention on threatened species.[22] To conform to this convention, the Monacan operation had to receive export permits, after consultation with national or international scientific authorities. At the same time, Doumenge had just been named president of the ecological commission of the International Union for the Conservation of Nature (IUCN).

In fact, shortly after the end of the Gulf War, the first expert evaluations had shown that the persistence and diversity of coral species and varieties had never really been threatened. A reporter for *Science et Vie* concluded as early as May 1991 that the heralded ecological catastrophe "appeared less grave than had been imagined."[23] Six months after the war, the quantity of hydrocarbons measured in the sediments on the

38   Saudi Arabian coast near Kuwait was equivalent to that observed on European coasts. The reduction in oil tanker traffic for several months had even contributed to a reduction in hydrocarbon pollution in the Gulf as a whole.[24] The polluted area was in fact limited to several coastal zones at the far end of the Gulf. As for the plan to save the corals, it posed numerous risks: transporting them could have led to elevated mortality rates, and mass culture of several dozen different species of corals, each with its own biological characteristics, was tricky.

To these concerns was added the fact that Jaubert, the director of the institution making the proposal, *also* held the patent on the culture system. Moreover, the director of the Oceanographic Museum of Monaco supported the project in the press,[25] doubtless favoring the import of corals because this would enrich the Museum aquaria and enhance their status. The project was rejected.

But while people dreamed of importing exotic corals to Monaco, another exotic tropical species, *Caulerpa taxifolia*, was establishing itself under the Museum and at Cap Martin. On this subject, from September 1990 to April 1991, the press was nearly silent. Only the fishermen of the prud'homie of Villefranche-sur-Mer–Beaulieu-sur-Mer–Saint-Jean-Cap-Ferrat, whose fishing territory was next to Monaco, had taken the initiative of publicizing their concern in the local newspaper.[26] Fishing nets had caught the alga west of Monaco at a depth of 180 meters.

A reporter from *L'Express*, Christophe Agnus, visiting Nice in March 1991 to prepare an article on the endangered corals of the Persian Gulf, wanted to meet me to get my opinion on the effects of hydrocarbon pollution on marine organisms. Never having conducted research on this subject, I did not hide my lack of expertise. However, I knew this type of pollution is reversible in the short or intermediate term, while the introduction of a dominant, invasive species is much more dangerous to ecosystems—it can become irreversible. I therefore mentioned the biological pollution constituted by the tropical alga introduced into the Mediterranean, suggesting it ought to be considered as posing a major risk for coastal marine biodiversity. This example was right here under our noses and museums. In hindsight, this analysis appears perspicacious: the spectacular damage caused by the hydrocarbons was in fact very localized; the burning oil wells were quickly extinguished. The Gulf corals remained alive, without having been transported to the Museum, while the tropical alga rapidly came to dominate and eliminate a part of the coastal marine flora of several hundred hectares on both sides of Monaco.

Surprised by this unusual story but convinced by my arguments, Agnus investigated the affair. Of course he showed up at Doumenge's office. While Doumenge probably expected to be queried only about the damage suffered by the ecosystems of the Persian Gulf, he was instead questioned about the little tropical alga that had "escaped" from the Museum. Did he want to retract his first, reassuring statements about the spread and persistence of the tropical alga? Did he know about the paper in press in the IFREMER journal and my appearances at the ministries to find the necessary funding to make a more specific assessment? Was not the proposal to culture corals in the Museum at risk of being thwarted by the revelations about the negligence in culturing the tropical alga in the Museum aquaria? He quickly changed his tune.

Whereas one month earlier (in February 1991), at a meeting held at the Oceanographic Museum on the creation of a reserve for dolphins, we had exchanged cordial and trite remarks on the problem of the alga, Doumenge now began a polemic by means of the popular media. The reporter noted with some surprise the remarks about me: "Meinesz plays the clown, the buffoon, in order to find funds for his laboratory." [27] Doumenge had found a motive for my interest in this alga.

I was advised to defend myself, to show outrage, to undertake a lawsuit. What good would it do? I thought that Doumenge's statements would sooner or later be discredited. I did not want to fall into the trap that was set for me—to exhaust myself emotionally and financially in a trivial judicial sparring match, which would cause our dispute to be seen as just another personal rivalry. His insults hardened my desire to strengthen and to publicize my assessment, based on detailed knowledge of *Caulerpa* and of the coastal ecosystems of the Mediterranean.

My administrative superiors should have been disturbed by these

40   declarations, very out of the ordinary in academic circles. A judge or a physician insulted in this way in the course of exercising duties would certainly have received the support of people in positions of ethical authority. On the contrary, in this case the intolerance of such remarks was more likely to make people smile: they preferred to remain on the sidelines, to display a "wise" neutrality, or to avoid the polemic by seeing both protagonists in the same bad light—both of us were wrong to have talked to reporters. People wished especially that I would drop this hazy story about an alga, which was above all threatening the institutional relations that some faculty of the University of Nice–Sophia Antipolis had cultivated with the Museum.

As for *Caulerpa taxifolia*, not only did Doumenge persist in asserting that the implantation of the alga was beneficial to the Mediterranean, but he declared for the first time that it probably did not come from the Museum. His declarations represented a fundamentally new stage in the "*Caulerpa taxifolia* affair," an attempt to exonerate the Museum. It is worth recalling and explaining them. Here are the arguments advanced by Doumenge:[28]

"The alga came from the Museum?"

"It's possible," replied François Doumenge, "but there are other scenarios: fouling (in which the plant would have crossed the oceans attached to the hull of a ship) or 'Lessepsianism' (in which it would have arrived via the Suez Canal). And who can even say that the alga is not a longstanding inhabitant of the Mediterranean, that it wasn't present in the last century? The especially mild winters that we have just had would have sufficed to allow the new growth."

His first hypothesis was fouling; the alga would have crossed oceans attached to a ship. The idea of a species attached to the hull of a boat arriving from the tropical Atlantic or the Red Sea, then detaching itself, by chance, just under the windows of the Museum, is a product of fiction, not science. First of all, *Caulerpa taxifolia* is rare in the regions where it is native. Further, the probability is infinitesimal that, on the one hand, it would be transported this way in the first place, and, on the other hand, it would detach in exactly this spot. One should know that the genus *Caulerpa* has never been recorded in the many publications that list species that foul boats (by attaching to their hulls). Moreover, anchoring boats is prohibited at the foot of the Museum because of the presence of the intake hoses for seawater for the aquaria. Finally, the alga had never been found in Monacan harbors.

The second hypothesis put forth was that of "Lessepsian" introduction, in which the species would have come from the Red Sea via the Suez Canal. Such species arrive on the Egyptian coast, in the eastern basin of the Mediterranean. The dominant currents carry them east, towards Israel and Lebanon. Waters laden with living organisms from the Red Sea are diluted in the many meanders of the Turkish archipelago and of Greece. These waters take a very long time to arrive at the boot of the Italian peninsula. Other currents coming from Gibraltar have a greater tendency to push these eastern waters back towards the east. When *Caulerpa taxifolia* first arrived, no current had yet carried algal species from the Red Sea to the northwestern Mediterranean. In all the countries bordering these seas, there are many oceanographic stations and excellent biologists and phycologists. The tropical alga *Caulerpa taxifolia*, very rare in the Red Sea, had never been noted until then on any Mediterranean coast. Nevertheless, Doumenge suggested that the alga came from the Red Sea and arrived straightaway in Monaco—fifty meters from the windows of the Museum and its tropical aquaria that contained, for several years, a completely identical stock of *Caulerpa taxifolia.*

Finally, Doumenge's third hypothesis was that it had been in the Mediterranean all along. In this scenario, the species would always have been in the Mediterranean and reappeared exactly in the territorial waters of Monaco (because the warm water rich in pollutants would serve as fertilizer for algae, as Doumenge subsequently hypothesized in other interviews).[29] We will return to this hypothesis, which testifies to a real scientific obscurantism.

Despite the implausibility of these hypotheses, Doumenge, echoed by a small group of scientific friends, reiterated them in many interviews during the four following years. The anecdotal suggestion of this mode of introduction of an alga neglected by the authorities was bound to become, by the virulence of his remarks, an interesting idea for the popular media. Three pages were devoted by *L'Express* to this scientific ecological subject,[30] which was rare for a major national news magazine. Although the article was generally favorable to my ideas, the publicized polemic opened by Doumenge led to another abrupt halt to the calm assessment of the problem.

By attacking me, Doumenge had succeeded in transforming an ecological problem into a quarrel between cliques. The reactions illustrate both the unease of observers and their allegiance to the stronger protagonist.

The anger of the Museum director had provoked such a great stir that many colleagues tried to convince me to avoid the sensitive subject of the origin of the alga. The urged me not to say, "the alga apparently introduced in 1984 under the Museum of Monaco," but instead, "the alga, which appeared at the beginning of the 1980s, was introduced in discarded aquarium material." It was more politically correct.

The article in *L'Express* appeared the day of my visit to Paris to present my research proposal to the ministries. The people I talked to had not yet read the weekly, but I imagine they did so later. Could the allusion, especially scurrilous, to the financing of my laboratory—while I was in fact soliciting "emergency battle funds" to sharpen the assessment—have been the reason the proposal was rejected?

Twenty days later, IFREMER produced, as we will see, an expert assessment of *Caulerpa taxifolia* without mentioning its origin.

Slander will always leave some traces. Faced with this virulent and unexpected opposition, I decided on a new phase of withdrawal, preferring to dive to the *Caulerpa* prairies to study them rather than to swim in the troubled waters of personal attacks and counterattacks. My proposals were in expert hands in the ministries, and I refused to get sucked into a futile debate.

May 1991: The French Research Institute for Marine Development Makes an Appearance

The proliferation of algae in the sea is a problem under the exclusive jurisdiction of the national government. Thus I waited for the authorities, alerted in writing by July 1990, to have the problem assessed by IFREMER, which has a consultative function for the Secretariat of State for the Sea and the Ministry of the Environment. But this institution had not been engaged. It did not manifest itself until then except in the press, where it attempted to calm the debate over the toxicity of the alga, a subject that had worried more than one fisherman. Thus, people could read twice in *Var-Matin:* "The aquaculture laboratory of IFREMER states that it [the alga] is eaten in the form of a salad or condiment in some countries of the South Pacific."[31] To assert that it is "eaten" was deceptive and, above all, had little to do with the essence of the problem posed by the algal invasion. Deceptive because, as was mentioned earlier, *Caulerpa taxifolia* is not edible and is so rare in the Pacific that most people of countries bordering the ocean would not even recognize it.

As for the ecological aspect, the researcher at the IFREMER laboratory at Nantes had advanced a surprising argument: "Moreover, *Caulerpa* is a green alga and thus contains much oxygen, which is, in fact, beneficial for the environment." Where the alga grew, there had in fact never been an oxygen deficit. In addition, *Caulerpa taxifolia* replaced other green, red, and brown algae that also contained chlorophyll and produced oxygen equally. These "clarifications," founded on specious arguments, seemed to me out of place in light of the risk that *Caulerpa taxifolia* posed to Mediterranean ecosystems.

Not until April 1991 did the departmental director of Maritime Affairs for Alpes-Maritimes officially engage IFREMER in the "phenomenon described by Professor Meinesz." I had authorized him to transmit my in-press paper to the director of the IFREMER branch at Toulon–La Seyne. The only response I received officially was a copy of a resulting opinion dated May 2, 1991. It had been formulated without a prior dialogue, as the bureaucrats of the Toulon–La Seyne IFREMER branch had not contacted me. But I was especially disappointed to observe that it had been drafted in an office in Toulon, without the involvement of IFREMER scientists or the external experts who had seen firsthand the invaded sites. The opinion was based solely on a reading of the literature. All the aspects related to on-site observations had been inspired by my study, which was not even cited. This opinion was not, in fact, an opposing view, but a simple interpretation of the situation I had described.

Nevertheless, I was at first satisfied, happy to have finally seen a document emerging from a well-known institute, concluding as I had that we should be concerned about this invasion of *Caulerpa taxifolia*. But after several rereadings, I was forced to conclude that this opinion was ambiguous. The author of the opinion thus declared: "It is extremely difficult, if not impossible, to fight such an invasion, and the fight will be difficult, long, and costly." Before adding: "This battle cannot be undertaken until the economic and nutritional problems are verified." The impact on ecosystems was not taken into account. IFREMER judged, however, that "the alga is destined to colonize the entire Mediterranean." If the conclusion underlined the gravity of the problem, the attitude of resignation prevailed over action—they affirmed that nothing could be done, all the while suggesting that they were not losing interest in the problem. In other words, if it was going to affect human health, we should act immediately; by contrast, if it represented only the potential overthrow of all Mediterranean ecosystems, it was not an urgent matter.

Since the submission of my manuscript to the IFREMER journal, my additional observations on the biology and spread of the alga gave me a bit of hope regarding the effectiveness of an eradication campaign. I began to understand that the invaded area was still limited to several hectares, mainly restricted to the vicinity of the point of introduction of the alga. The arguments that it was hopeless, again made without any field observations, seemed unjustified to me.

Yet another surprising element, discussion of the origin of *Caulerpa*   **45**
taxifolia, was dodged altogether. Only one timid allusion could be found,
in one of the last paragraphs, which again took up the recommendations
of my report, but in doing so eliminated all reference to the aquaria of
the Monacan Museum: "It is likely that this is not, unfortunately, the
last we will see of accidental introductions. Thus, an information campaign should be undertaken to sensitize the public, especially persons
likely to cause such introductions."

The Institute adopted a compromise between Doumenge's hypotheses on the origin of the algal introduction, hypotheses defended in the
L'Express article that had appeared several weeks earlier, and my scientific publication. This diplomatic amnesia about the site of its first appearance, of its most likely origin, of the characteristics of its dissemination from that point, certainly helped to discourage any real grasp of
the pending crisis. Without these facts, the appearance of the alga could
appear as a natural, uncontrollable, universal, inescapable phenomenon.

In spite of everything, IFREMER prudently advised the director of
Maritime Affairs of Alpes-Maritimes to "keep tabs" on the phenomenon. This recommendation was wishful thinking. This ranking bureaucrat had done his job—he had enlisted the competent administrative
authority to produce an assessment. The administrator of IFREMER
had "thrown the ball back to his court," knowing full well that the departmental director of Maritime Affairs had neither the means nor the
expertise to undertake the relevant research.

As its title indicates, the French Research Institute for Marine Development (IFREMER) is an organization whose mission is, above all,
to exploit marine resources.[32] Its recent involvement in marine environmental matters made the Institute the final authority in this affair. Now,
though the new IFREMER laboratories for coastal environmental research could use substantial means (oceanographic vessels) to explore
the continental shelf, they were ill-equipped to investigate small soft-
bottom patches (three licensed scientific divers for the entire Mediterranean). By the same token, their resident experts were more competent
in aquaculture or in measuring anthropogenic pollutants than in marine

46 ecology. This partly explains why the entity charged with supervising the management of the marine environment did not undertake either research or an in situ assessment of the problem.

July–August 1991: First Support from Scientists

The fruitless calls for help, the public debate focused on a scientific pseudo-controversy, the attacks by the director of the Museum, and the pseudo-expert assessment of IFREMER had certainly all bothered me. But they had once again strengthened my resolve. In order to proceed, I was going to have to find impartial scientific collaborators.

In the spring of 1991, I sought in vain among my colleagues in chemistry at the University of Nice for someone who would be able to perform, gratis, urgent research on the toxins that might occur in the tropical alga. No one was available. In fact, all had to honor research contracts for their laboratories. Serendipitously, I received at this time a call from a professor of organic chemistry at the University of Trent, in Italy, who had learned of the alga thanks to the article in *L'Express*. A specialist in the extraction and identification of toxins concentrated in certain marine species,[33] Francesco Pietra suggested, without any demand for funding, that he would work on the *Caulerpa* toxins. This spontaneous and fortunate collaboration was very welcome. At the beginning of July 1991 I collected, in one dive, fifty kilograms of *Caulerpa* at Cap Martin. Early in the morning, I carried the algae 250 kilometers from Nice to a factory that could freeze-dry them. It was with these dried specimens that Pietra began his research. Before the autumn, he informed me of the first results of his analyses. I was not surprised when he disclosed that terpenoids (a class of lipids, many of which are toxic) were present in significant quantity in the alga.

For his part, my Marseilles colleague Boudouresque was increasingly convinced by my arguments. Despite the lack of research funding to work on *Caulerpa*, despite the lack of interest of the ministries that

had been contacted and the unpleasant atmosphere created by the polemic, he decided to help me. He suggested undertaking an experiment
on the feeding behavior of one of the main herbivores of the Mediterranean, the edible sea urchin *Paracentrotus lividus*. He prepared four
aquaria in a closed circuit. Ten urchins were used in each tank. In two
tanks, he provided algae favored by the urchin, and in two others he
placed *Caulerpa taxifolia*. The weight of the algae consumed and the
daily volume of matter excreted by the urchins were measured every
other day. After three weeks, Rodolphe Lemée and Xavier Mari, students in my laboratory, took over the experiment, which ran for over
two more months.

A simple test allowed an assessment of the vitality of the urchins.
If they are turned upside down, they deploy dozens of tiny suckers that
they attach on one side in order to haul themselves over to regain their
natural position. Urchins generally somersault in this way in less than
ninety seconds. Now, if they are in bad physiological condition (exposed
to pollutants, malnourished), they take more time to turn over (up to
fifteen minutes). To evaluate the effect of a diet based on *Caulerpa* on
the vitality of urchins, Boudouresque and my students timed the somer-
saults of the urchins every three days. The experiment was convincing:
during the summer and autumn, the urchins ate little during the first
weeks, then let themselves die of hunger rather than eat *Caulerpa taxifolia*. Appearing healthy, they turned over more and more slowly. By
contrast, the control urchins, fed the local alga, were in great shape despite their captivity. Other urchins, fed a mixture of Mediterranean algae and *Caulerpa*, studiously avoided eating the tropical alga. These preliminary results accorded with the literature survey reported in my
scientific publication. Thus, as in the tropics, these marine animals eat
no (or very little) *Caulerpa*. This also confirmed our diving observations:
the local fauna do not eat this alga. These toxins are repellents.

Boudouresque's experiments tended to confirm that the algal invasion was worth worrying about. The implications were easy to understand. On the one hand, the alga had obviously become a dominant

48 competitor of the other marine plants of the Mediterranean, and, on the other hand, it provided little or no nourishment to the local herbivorous fauna, which would thus be threatened in turn, sooner or later.

In joining the assessment and confirming my hypotheses, Pietra and Boudouresque began to understand what was at stake. I was no longer on my own. This was the first time since the beginning of this story that other specialists had seriously assessed the impact of the algal invasion.

August–September 1991: An Attempt to Minimize the Problem

On the initiative of a reporter from the western part of the Department of Var, two articles on *Caulerpa taxifolia* appeared simultaneously on August 5, 1991 in *Var-Matin* and *Le Provençal*, respectively. Boudouresque had just begun, in close collaboration with my laboratory, his first work on *Caulerpa*, and he had publicized this fact. Reporters from *Agence France-Presse* queried the various players in this drama. A dispatch that was circulated on August 7, 1991 was widely reprinted by editorial offices in a period with little compelling news, and during a season when millions of tourists at the seashore were more receptive than normal to anything about the marine environment.

About ten articles appeared in the local and national press. One can read the deliberately alarmist headlines, which often appeared on the front page. First on the danger (from August 5–7, 1991): "Green threat to the deep blue sea,"[34] "An alga smothers the Mediterranean,"[35] "Beautiful but poisonous,"[36] "The alga that threatens the Mediterranean,"[37] "The cursed alga smothers everything under 8,000 leaves per square meter."[38] Then on the polemic (beginning August 8, 1991): "Ecological catastrophe for some, salubrious recolonization for others. Polemic on the proliferation of algae in the Mediterranean,"[39] "The alga that sows dissension, specialists are divided on its harmfulness,"[40] "*Caulerpa taxifolia*, toxic or useful?"[41]

Duly contacted, Doumenge had rekindled interest in the subject by his contradictory opinions and his mockery. A review of the newspaper articles showed that he essentially advanced three arguments:

He rejected the responsibility of the Museum in the algal introduction, largely by confounding the Mediterranean *Caulerpa* species and *Caulerpa taxifolia.* This was the hypothesis that it had always been there, already presented in *L'Express.* After having cast doubt on the identification of *taxifolia,* indeed even on *Caulerpa,*[42] he increased the confusion with the Mediterranean *Caulerpa (prolifera).* In *Le Monde,*[43] he evoked the good points of *Caulerpa prolifera,* which had unfortunately declined because of pollution.[44] To convince those who did not understand the subtleties of marine salads, he stated that *Caulerpa taxifolia* had always existed in the Mediterranean, disguised in another form. Thanks to the mild winters, it was transformed into the beautiful *Caulerpa taxifolia.*[45] This argument left nonspecialists believing that *Caulerpa taxifolia-prolifera* was Mediterranean and that its growth was not out of the ordinary. Because it was natural, we could not do anything about it, and, above all, it did not come from the Museum! Unfortunately for the Museum, this spontaneous appearance arose just fifty meters from aquaria containing *"taxifolia."*

Other peculiar arguments were advanced to exonerate the Museum: "If it came from our aquaria, and all of ours, at present, function as a closed system, it would have been decimated by the thermal shock! The water of the Mediterranean is, after a decade of mild winters, 13°C maximum, and that of the aquaria is 28°C."[46] However, these aquaria were not all in a closed system, and Doumenge knew that the alga could have been implanted in the summer, a season when the water of the Mediterranean sometimes reaches 27°C: thus no thermal shock would have occurred when the *Caulerpa* hit the water.

Doumenge also bragged about the usefulness of the alga in enriching the seafloor, which he felt was nothing more than putrid mud. Here, he saw the invasion of the alga as so beneficial that he almost claimed credit for it. He often called it *taxifolia* in this context: *"Caul-*

erpa taxifolia is implanted in muddy zones that swarm with life today. This is beneficial."[47] "In beds of mud laden with organic matter, a strongly eutrophicated environment, the algae are now encountering a very rich milieu and are growing spectacularly. This is a good sign."[48] "*Caulerpa?* A real blessing! Previously under those boulders there was only mud. Today there is a luxuriant prairie."[49] "It is beneficial for our sea because it is part of the recolonization of heavily polluted zones. We have veritable green prairies instead of muddy plains."[50]

The argument advanced first by IFREMER on the properties of the alga in oxygenating water was also recycled: "Listen, these algae, they oxygenate the water, since they're green."[51] To complete the inventory of benefits conferred by the alga, the director of the Museum emphasized how aquarists felt about it: "*Caulerpa taxifolia,* the aquarists love it, which is the perfect proof that it is not damaging."[52] As if one could not love an alga—appreciate its decorative qualities and tolerance of an aquarium environment—and at the same time be worried about its proliferation in the open sea!

To perfect his argument, Doumenge had to attribute a motive to the alarmists. Their pointless ecological concern, was it not primarily generated by crass economic motives? To hear him speak, *Caulerpa* was nothing but a "sea serpent" that surfaced every year in August.[53] On television, the director of the Museum said reassuringly, "It's a sea serpent, the only good thing about it is that it makes people talk about the sea and that makes money flow to the laboratories."[54] "All this, it's staged! He [speaking of me] is trying to get money for his laboratory!"[55]

This was a particularly deceptive presentation of the problem: it insinuated that I had fabricated the entire business out of whole cloth to fill the coffers of my laboratory, when in fact I basically needed support to extend our knowledge of the introduced alga. In the sumptuous leather armchair of Prince Albert I of Monaco, invested with the heritage of Commander Cousteau, he repudiated and discredited a modest "prof" of a provincial "school" who had been impertinent enough to contradict him. His attacks did not affect me. In fact, he misidentified his adversary: he was playing against nature, against an alga that, far from the above-water debates, was growing tranquilly and inexorably.

In this unpleasant context, I resolutely continued my research and again opened contacts with the authorities of the marine domain. I responded to reporters who sought me out on their own initiative. To limit the risk of hasty or excited interpretations of my statements, I always asked them to refer to my scientific publication. I forbade myself to comment on the personal attacks hurled at me. An aggressive attitude was not in my nature; I was content to relate the facts described in my article and to suggest scenarios of how things might develop. In all my interviews, and until July 1992, I never said the name of the director of the Museum. Actually, it was the fact that I stuck to my guns that unleashed Doumenge's ire.

Curious about these "conflicts," *Agence France-Presse* (AFP) took the initiative of questioning IFREMER. François Madelain, its spokesman and director "of environmental matters" at the Parisian headquarters, had declared to AFP, "IFREMER wants to calm the polemic that has developed over the effects of the rapid growth of a tropical alga. . . . The prudent path is for IFREMER to stand aside from this polemic and to serve as a mediator, and to stand by this statement by François Madelain: This is not the first case of the introduction of a foreign alga into our waters; we have observed exactly the same phenomenon in the Atlantic, with *Sargassum muticum*, which arrived with oyster cultures from Japan several years ago and which replaced native species. In the case of *Caulerpa*, we have not observed ecological catastrophes or any impact on the fauna. It is important not to dramatize the situation. The results of our analyses have moreover proven that no fish can be poisoned by *Caulerpa*."[56]

This stance confirmed how Doumenge's intervention had been taken into account by the central administration. His media statements were officially countered by my study, which was then in the course of being published in a scientific journal. That such angry statements excited the interest of people who loved scoops and, by extension, that of

52   the public at large, is understandable. But it was shocking that these supposedly scientific declarations and the various calumnies reported in the news media were given the same credence as careful studies and observations. However, this was the case; the spokesman for IFREMER had not hesitated to locate himself between the two viewpoints. Worse, instead of keeping a strict neutrality and studying the phenomenon firsthand, the arbiter leaned towards the view of the polemicist, in that he minimized the problem. This behavior well illustrates the evolution of credibility in scientific communication: for journalists, politicians, and even government experts, news spread by the media is now considered on a par with scientific publications.

"We have not observed. . . . The results of our analyses have moreover proven." The most disconcerting aspect of this statement was to learn that IFREMER intimated that it was involved in studying the problem, when, in fact, no bureaucrat of that organization had ever seen the alga. Following Doumenge's lead, the spokesperson for IFREMER downplayed the situation and, in so doing, misinformed the public. I tried to reach him the day extracts of the dispatch appeared in several newspapers. I was told he had just left on vacation. I could have requested a clarification in the newspapers that had picked up the AFP dispatch, but it was useless to throw oil on the fire. A month later, the director of IFREMER confirmed to me that his institution had undertaken no research or analysis, and he attributed the false information to misinterpretation by reporters.[57] However, two reporters who had questioned the spokesperson were adamant: they had correctly cited his statements without changing them.

Still in August 1991 the IFREMER branch at Toulon–La Seyne was questioned by Roger Cans, reporter for _Le Monde_. Cans met with an expert on anthropogenic pollution who had seen the alga only on television or in the local newspapers; this person had picked up some excerpts from my publication and appeared reassuring. In _Le Monde_ of August 14, 1991 one could read: "At the IFREMER center at Toulon, they are much calmer." I met Cans several months later by chance in Corsica, and I reproached him for not having asked the question: have you seen the alga? The reporter sincerely thought that IFREMER had studied the problem; it seemed so obvious to him! Although the person he had interviewed had not stated any falsehoods and had even confirmed the risks associated with the toxicity of the alga, I was surprised to see an expert issuing pronouncements on an ecological problem that he had not observed himself. It may have been possible to be calm only from this vantage point.

Cans disappointed me deeply by his analyses of this matter. Over the next two years, I met him three times at meetings on marine reserves, without the opportunity to tell him my opinion on the *Caulerpa* invasion. He asked me rather sarcastic questions; his opinion seemed fixed. He was firmly convinced that this whole *Caulerpa* story was a set-up; his articles often tended to sow doubt on the problem and to reinforce the polemic.

In 1992 he showed his true colors in his pamphlet *Tous verts.*[58] His conclusion unpleasantly reminded me of Doumenge's interpretation of this affair: "If one succeeds in persuading the government that it's a question of life or death for the Mediterranean, one can hope for exceptional research funds. It's a good bet that it's still a matter of a warning that's half a settling of a score, half an appeal for research funding." He even added another hypothesis: "In fact, it happens that it is the Oceanographic Museum of Monaco, longstanding rival of the labs of the University of Nice, that is accused of having introduced the alga from its aquaria." If he had visited our mini-laboratory of 110 square meters, he would never have thought to impute to us any sort of sense of rivalry with the gigantic and prestigious Oceanographic Museum. Moreover, I had always had good relations with the research laboratories of the Museum. The director of my former laboratory at the University of Nice, Raymond Vaissière, had long served as an official of the Museum.

Initially, I had had only a single opponent, Doumenge. He was an adversary of stature because he held the prestigious posts of director of the Oceanographic Museum of Monaco, president of the environmental commission of the International Union for the Conservation of Nature (IUCN), and secretary-general of the International Commission for the

54   Scientific Exploration of the Mediterranean (CIESM). Some journalists were doubtless impressed by his prestigious titles.

At the end of summer 1991, I observed that all the spokesmen of IFREMER had politely followed his lead in minimizing the problems. I was dismayed by such a facile attitude on the part of government experts and high-ranking administrators who gave statements on this affair; they used their dominant positions or the power of their institution to confer authority on their statements. Given the media debacle, given the doubt, why had they not attempted to verify my observations, why had they not sent an expert to do so? This was a bad period for me, but I was convinced that time would play in my favor. The situation was crystal clear—no alarming observation of *Caulerpa* had been discredited. There was no well-founded criticism of my report. By contrast, my stand was upsetting people. And all the while, the alga grew.

Autumn 1991: A Public Meeting against the Indifference of the State

By the beginning of 1991, my collaborators and I had dived on many occasions at all the invaded sites and to the new colonies found by our informants. We had gone over several dozen underwater kilometers from Menton to Monaco using a dive-scooter.[59] Another method, underwater towing, was used for rapid exploration of the small patches of soft bottom situated west of Monaco. In this approach, a diver holds a board furnished with two handles, which is pulled by a boat. Underwater, one sees the bottom pass rapidly by, and one can easily ascend, descend, or let go. In this fashion, we had ourselves towed for several kilometers by the boat of the General Council of Alpes-Maritimes.

In autumn 1991 I had a good overall appreciation of the invaded area. By comparison with the situation at sites visited one year earlier, I was better able to understand the dynamic of the invasion. In sites where, only a year earlier, we had found a dozen very sparse clumps of *Caulerpa*, a continuous lawn was now established. *Caulerpa* had progressively penetrated the *Posidonia* meadows. It was clear that the growing competition between the two plants threatened the structure of the most representative ecosystem on small soft-bottom patches of the Mediterranean. The spectacle of these bottom patches fouled by *Caulerpa* was so distressing that, with each resurfacing, we were increasingly disturbed by the crushing above-water indifference. Until December 1991 the members of my laboratory were the only scientists to have seen *Caulerpa* invade Cap Martin. It was obvious, however, that a single dive was worth more than any argument.

That autumn, the alga was west of Monaco, at Cap d'Ail, but also much further, in front of l'Estérel, on the coast at the town of Saint-Raphaël (Agay) and in front of the Maures Mountains at Lavandou, opposite the Port-Cros National Park. At Toulon, the area occupied by the alga had grown tenfold in one year. To the east, we were notified of its presence in the port of Menton and in front of the customs post on the Italian border. We verified and mapped all new colonies.

The evidence for the spread of the alga piled up. Nevertheless, from April 1991, all the approaches to relevant ministries had failed. The research arm of the Ministry of the Environment, petitioned by mail on two occasions, remained silent. I had received two letters from the Interministerial Committee for the Sea. In the first, a letter of June 16, 1991, I was asked to be prepared to respond to additional questions that would be asked of me by a technical center of phycology, CEVA.[60] Located at Pleubian in Brittany, this institution was far from the sites invaded by *Caulerpa taxifolia.* Its interests were also very different from ours: its role was above all to promote the use of algae for food or for the production of gelling agents. A second letter, of September 6, 1991, contained several trivial questions posed by a CEVA phycologist. I was notified at the same time that the Secretariat of State for the Sea could not fund the proposal for new research. We were led to hope for eventual support from CEVA. Reached by telephone, the director of that institute did not think it at all possible to become involved in that kind of research. The Secretariat of State for the Sea obviously wanted to put me on hold.

While it became increasingly urgent to consider our assessment if

56   there was to be any hope of controlling the phenomenon, nothing had
made us optimistic that the problem would be quickly confronted.

Our reports and hypotheses on the changing situation remained
contested by the powerful director of the Museum. His hostile and aggressive attitude precluded any collaboration or calm scientific debate.
The communication of my study to the scientific community by means
of the IFREMER journal had elicited only a few encouraging remarks,
and IFREMER ostensibly minimized the problem. The reasons for their
lack of interest in the problem, more political than scientific, remained
obscure. The press statements by their spokesman left no doubt as to
their partisan, wait-and-see strategy. The relevant ministries, influenced
by the IFREMER experts, remained deaf to our pleas, or at most
slightly concerned, thus blocking assistance at the decision-making
level. The presentation of the problem by the media had remained mar-
ginal, essentially left to the initiative of a few reporters. The minimizers
used the only means of communication available to them—the media—to propagate their reassuring hypotheses.

In this context, it was necessary to act as quickly as possible. To
revive the debate, I took the initiative of organizing a meeting where
scientists would be able to dialogue, to present the problem to the
affected government agencies and associations, all in the presence of the
press. I was finally going to be able to divulge the idea that had dominated me for several months—the potential eradication of the alga! I
had gradually changed my views on this topic. At the end of 1990, after
drafting the article for *Oceanologica Acta*, I believed that eradication
was impossible. I thought at that time that I was seeing only a small
part of the invasion. How many little colonies had been able to establish
themselves, and where? No one could answer these questions. But the
data gathered on the invaded sites during the past year and the observations that reached me, as much by records of new colonies as by the
absence of any algae in vast areas, had given me a better idea of the
range, the speed, and the mode of dissemination of *Caulerpa taxifolia*.
With this information in mind, I began to feel that eradication of *Caulerpa taxifolia* might be possible.

The first premonitory signs of sexual reproduction were observed at the end of July 1991. Certain *Caulerpa* individuals presented fronds of a different color from others, and these were covered with papillae. Inside the plant, instead of the usual organelles (nuclei, chloroplasts), could be seen numerous spherical corpuscles containing chloroplasts. These correspond to the first stages in the formation of male or female gametes. However, I was unable to observe the final evolution of the very ephemeral sexual cellules and therefore had yet to confirm this mode of reproduction. If *Caulerpa* reproduced sexually, it must have done so quite ineffectively. The spread remained quite limited; the alga was, at this time, quite restricted on either side of Monaco. The eggs (zygotes) arising from sexual reproduction could not have been the basis for a generalized, rapid dispersal over hundreds of kilometers of coast. In fact, only three distant implantations had been found at this time (on the Var coast, at Toulon, Lavandou, and Saint-Raphaël); these were recent, as the occupied area totaled less than 100 square meters. Accidental cuttings, caused by humans (fragments transported by systems of anchorage of leisure craft mooring in great number over the lawns of *Caulerpa* at Cap Martin) could have been the origin of these implantations.

The spread of the alga was not very rapid except around the points that had been invaded first. The bigger the area of a colony, the greater the number and dispersion of the small secondary colonies. In a contaminated site, the speed of the algal spread was proportional to the mass of algae already present. Cuttings that were torn off did not float; they sank and thus could not establish except in the immediate vicinity of their origin. By the same token, I did not think that the eggs would be carried by currents but rather would land around the site of their production. Cuttings and eggs could explain the presence of many new colonies around the oldest ones.

In the autumn of 1991, before the quiescent season for *Caulerpa taxifolia*, we had estimated the area more or less invaded by the alga.[61] In a year it had grown from three hectares to around thirty hectares, the equivalent of nearly sixty soccer stadiums. This area was distributed

58   almost wholly on both sides of the site of introduction (Monaco), along
only three kilometers of coast.

The eradication of the alga over such an area, either manually or
chemically, would have without any doubt halted the process of spread.
The small fragments that would have been missed and would have
grown in the following months would have been easily noticed and
could have been eliminated before they underwent further reproduction. Repeating the operation several times in a row, we would have
been able to end the problem. Attempts at manual eradication, although
very laborious and slow (two square meters cleaned on average by two
divers in one hour), allowed me to conceive of a successful eradication
program, on the condition that many divers were mobilized for several
weeks and above all that the effort begin very soon. With more effective
means and the assistance of biologist divers, other techniques, even ones
that might put a lot of pressure on the habitat to be treated, could also
be envisaged. For the small, distant colonies, I advocated the creation of
a team of three divers furnished with light eradication equipment. In
less than two months, they would have been able to eradicate the three
infestations farthest from Monaco (Saint-Raphaël–Agay, Toulon, and
Lavandou), an area totaling less than one-hundred square meters, then
maintain surveillance over the sites to eliminate any cutting that had
been missed. One thing was certain—we had to act quickly. The alga
had re-entered its winter period of vegetative repose. In six months, it
would proliferate anew.

We had presented our eradication strategy for the first time in a
report on the status of the situation, prepared to wind up the contract
with the General Council of Alpes-Maritimes and the Regional Council
of Provence–Alpes–Côte d'Azur.[62] The localization of all the colonies
and their progression since the beginning of 1991 justified the eradication. We emphasized the importance of timing; any delays would reduce
chances of success and increase the costs of the operation, regardless
of the chosen techniques. I nevertheless remained realistic, assessing
the chances of success as weak and recognizing that such an operation
would be very onerous. But we wanted to persuade the decision-makers

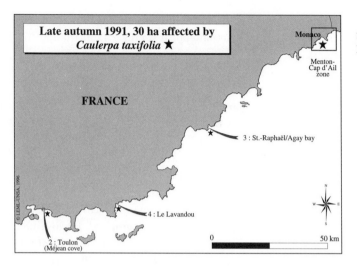

Late autumn 1991, 30 ha affected by *Caulerpa taxifolia* ★

Monaco ★

Menton-Cap d'Ail zone

FRANCE

3 : St.-Raphaël/Agay bay

4 : Le Lavandou

2 : Toulon (Méjean cove)

© LEML-UNSA, 1996

0   50 km

FIGURE 2.1.
Late Autumn 1991

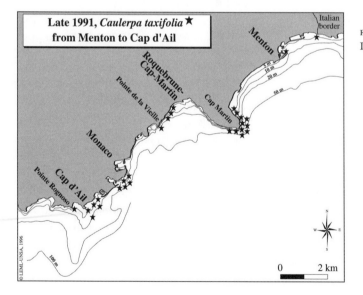

Late 1991, *Caulerpa taxifolia* ★
from Menton to Cap d'Ail

Italian border

Menton

Roquebrune-Cap-Martin

Pointe de la Vieille

Cap Martin

Monaco

Cap d'Ail

Pointe Rognoso

5 m
10 m
20 m
50 m

100 m

© LEML-UNSA, 1996

0   2 km

FIGURE 2.2.
Late 1991

60   to try it, because the confirmation of the invasion and alteration of marine ecosystems would make inaction appear a graver fault than a vain
attempt to stop it.

This report, accompanied by an invitation to the meeting, was sent
in November 1991 to the General Council and to the Prefect of Alpes-
Maritimes, to the Regional Council of Provence–Alpes–Côte d'Azur, to
the relevant ministries (Ministry of the Environment and Secretariat of
State for the Sea), and to the Prince of Monaco. In the following days, I
received a very terse letter from the Interministerial Committee for the
Sea.[63] The director of research recalled his previous action, confirmed
that he could not subsidize research on this subject, and felt that it was
up to the center for the study of algae at Pleubian in Brittany (CEVA)
to decide on the priority to assign to our research proposal. He declined
our invitation, feeling that a visit to our laboratory by a member of the
Interministerial Committee for the Sea could not "contribute to the proposed research." It was clear that the Secretariat of State for the Sea did
not want to hear any more about the alga.

A week after the invitations were sent, I received a call from the
Ministry of the Environment. A bureaucrat from IFREMER, assigned
to the ministry, asked me to fax him a copy of the entire file I had submitted in April 1991, as it could not be found. I understood why my
letters attempting to rekindle interest in the file had remained unanswered. For once, this new action was followed by a rapid reaction, but
it was still frustrating.[64] After questions very far removed from the
matter at hand (on the relationship of the alga to the carbon cycle or
the toxicity of herbivorous fishes in the Gulf of Gabès in Tunisia), the
IFREMER bureaucrat asked me what actions the Secretariat of State
for the Sea intended to undertake relative to this affair. Thus the representative of the Ministry of the Environment was asking me about the
thoughts of other bureaucrats with whom he was going to rub shoulders
in the Interministerial Committee for the Sea. I had the impression that
the bureaucrats of the two ministries were applying a schoolboy strategy
towards potentially embarrassing classmates—respond to them with
questions and send them to someone else.

At the end of November 1991, I asked the prefect of Alpes-Maritimes for a meeting to discuss the situation. I sent him my publication and a letter asking him to study immediately the feasibility of eradicating the alga. I also noted the usefulness of prohibiting the sale and culturing of the alga in public and private aquaria.[65] Finally, I emphasized the urgency of both of these measures, while reminding him of the response of the government when the oil tanker *Haven* ran aground in the Gulf of Genoa in the spring of 1991 (they were able to mobilize many people and, on the spot, to free up all the funding necessary to reduce the pollution). The prefect received me on December 9, 1991; he explained to me that he had not responded to my letter of the previous July because of the outcry raised by the Museum (I did not have sufficient credibility, in his eyes). He promised that he would ensure that the problem was studied; I subsequently received no indication of his further engagement in this matter.

On December 4, 1991, the director of the environmental television show "Sauve qui veut"[66] selected me as "man of the month." I was supposed to present the research of my laboratory on the restoration of Mediterranean bottom patches by transplanting *Posidonia*. I managed to convince the master of ceremonies of the show, Henri Sannier, to speak for two minutes about *Caulerpa*. I was therefore able to show several pictures of invaded sites, and I took advantage of this live broadcast to announce the meeting scheduled for December 16, 1991. In the hallway, I met the Minister of the Environment, Brice Lalonde, who was present for the telecast (which highlighted waste treatment). In five minutes, I presented the algal problem to him; he did not know anything about the story, but he promised to discuss it with his staff.

Four days later, Lucien Chabason, advisor to the minister, contacted me. I briefly described my previous disappointing interactions with personnel of his ministry, who had possessed all the information on the case for the last eight months. He assured me that he would defend financing the preliminary studies to ministry staff. This was a long-awaited sign of encouragement.

Meanwhile, I had convinced my colleague Boudouresque to promote the idea of eradication. At the conclusion of a meeting of the scientific committee of the Port-Cros National Park held in late 1991, he informed the press of our strategy. He even mentioned chemical eradication of the alga.[67] He sensitized the members of his committee, who were unanimous in their disturbed responses to this worrisome invasion. His steps were well publicized locally.[68]

Organizing the December 16 conference was laborious. Because all the large lecture halls of the faculty of sciences of the University of Nice–Sophia Antipolis were fully booked from morning to night, I asked the university president if we could use the theater of the "Château de Valrose." This castle, university administration headquarters, located opposite our research building, is a prestigious site that would have conferred a degree of solemnity on the meeting. To support my request, I sent my scientific publication to the president. He was a biologist and able to understand what was at stake. But he had never consulted me on the historical background of this affair and did not understand that we had exhausted all official avenues to make ourselves heard. I received his belated, negative response; in his letter, he deplored the press presence at the meeting.[69] The university did not want to damage its relationship with the Oceanographic Museum of Monaco. Jaubert and Doumenge were held in high regard by many university colleagues.

I skirted this obstacle so that the meeting could be held. I would long be criticized for this initiative; I had infringed on the obligatory reserve of federal bureaucrats. In their eyes, the main thing I had done was to perturb their skillfully maintained equilibrium. In departing from the routine of teaching researcher, I found that there was no structure facilitating dialogue, no place for a discussion of the ethics that ground research institutions and universities. Fundamentally, the university hierarchy was content to read the local newspapers and did not wish to hear me.

Fortunately, I had good relations with certain local elected officials. At the beginning of December, a municipal councilwoman of the city of Nice,[70] who had previously supported me in my opposition to the proliferation of yachting harbors, spontaneously offered me the use of a magnificent lecture hall located on the Promenade des Anglais. To pay the various costs of the meeting, I still had some funds left from contracts for the publicity campaign.[71] Two Italians agreed to participate in the meeting: Francesco Pietra (who had just finished his analysis of *Caulerpa taxifolia* toxins) and Francesco Cinelli, phycologist at the University of Pisa, who was well known in Mediterranean scientific diving circles.

Of course, all the researchers from the IFREMER station near Toulon–La Seyne had been invited, as had Thomas Belsher, responsible for studying the invasion of another alga in the Atlantic. Based at Brest, he was the only IFREMER phycologist who was a certified diver, and he knew the Mediterranean well. We had collaborated on several research projects mapping submersed vegetation in the Mediterranean and Polynesia. His name was on the list of experts consulted on the problem of *Caulerpa taxifolia* in May 1991 by the Toulon–La Seyne branch of IFREMER. I hoped that he would be able to inform his superiors of the state of the situation.

Before the afternoon meeting, I suggested that all these oceanographic specialists dive to see with their own eyes the impressive algal invasion. At daybreak the weather was splendid. Boudouresque and Verlaque, of the faculty of science of the University of Marseilles-Luminy, Cinelli and two of his collaborators, Pietra, two divers from the Port-Cros National Park, a team from Channel 2, an underwater photographer from *Figaro* magazine, and local journalists were all present, but there was no representative of IFREMER.

Belsher was supposed to come, but he informed me two days earlier that he was backing out. Very embarrassed, he told me that he had just met, by accident, his administrative superior, the director of the environment for IFREMER, Alain Merckelbagh, who wanted only the local representative of IFREMER, Yves Henocque, to attend the meeting. Henocque was a newcomer to the coast, having formerly been assigned to the Paris headquarters of the institute in an administrative position concerning international relations. He had a doctorate in biology and had studied the growth of shellfish and fishes in aquaculture. On the coast, Henocque found good company. He shared with Doumenge strong ties to Japan. Both spoke Japanese, both were members of the governing board of the same French-Japanese oceanographic society and had participated in several study assignments in Japan. The assistant director of the Museum of Monaco, François Simard, had just been recruited by Doumenge among the swarm of young researchers in the French-Japanese oceanographic society. Several years earlier, Henocque and Simard had organized a colloquium for the society. During the three following years, Henocque, promoted to director of the IFREMER laboratory of the Mediterranean environment at Toulon–La Seyne,[72] handled the *Caulerpa* affair locally for his administrative superior, Merckelbagh.[73] Although a diver, Henocque was not yet authorized to dive for the institute. As there was no other diver at the IFREMER branch at Toulon–La Seyne, no one from that institute was able to use this opportunity to observe the impact of *Caulerpa* in the Mediterranean environment. Merckelbagh had thus precluded all involvement of his institute in the visual inspection of the phenomenon.

The president of the dive club of Roquebrune–Cap Martin, distressed to see his favorite diving sites fouled by *Caulerpa*, did everything in his power to help me. He opened the clubhouse to us, placed the club outboard motorboat at our disposal, and two of his friends came with small personal boats. Three boats overflowing with men and materiel thus embarked for Cap Martin. Underwater, I imagined the astonishment of all the divers. *Caulerpa* was at the height of its seasonal growth and was most impressive. I was happy to have cornered so many experts, happy to have benefited from the splendid weather and the assistance of the dive club. I was certain that people were finally going to listen to me.

As soon as they emerged from the water, Boudouresque and Cinelli declared that they had just witnessed a frightening, abominable situation—it was a catastrophe! They made extensive comments on the impacts they had observed; the alga was beginning to cover the *Posidonia* seagrass meadows, there were almost no other species of algae on the invaded boulders, it was incredible! Every possible superlative was used.

In the space of a half-hour dive, the skeptics and the people who had been influenced by Doumenge's statements suddenly realized that *Caulerpa* was growing far beyond the putrid, sterile mud.

The meeting was held in the afternoon, in the presence of fifty representatives of local citizens' groups, environmental organizations, and fishermen's associations. The scientists showed the preliminary results of their "clandestine" research, none of which was officially supported financially (the necessary funding had been bootlegged off other research programs). Pietra revealed the nature of the toxic substances that he had just identified in Mediterranean *Caulerpa taxifolia;* they carried a lot of toxins, including new ones that had not been described until then. Boudouresque presented his work on the sea urchins, who starve to death rather than eat the alga. Verlaque, research representative of the CNRS, spoke about risks associated with introducing species into the marine environment. Pesando, from INSERM, discussed her research program, which was held up by the ministries for eight months; she demonstrated the utility of toxicological tests that would be suitable for the tropical *Caulerpa.* For my part, I shared my observations on the spread and exponential growth of the alga.

In conclusion, the scientists all held the same conviction—the alga is harmful to the coastal environment and, because of its rapid advance, its dominance, and its toxicity, constitutes a sufficiently worrisome risk that its eradication should be envisioned. They called for the urgent study of eradication techniques and an assessment of their costs. To coordinate our efforts, a scientific committee was formed, with me as president. Participating in the discussion, IFREMER, represented by Henocque, tried to make the situation seem less dramatic. As had the four previous spokespersons for IFREMER in this matter, he minimized the toxicity and ignored the impact on ecosystems. He proved very skeptical about the most widely suspected origin of the invasion, favoring some hypotheses raised in *L'Express* by Doumenge. He is not a phycologist and his speech was punctuated with gross errors that the specialists in attendance noticed with consternation.

The strategy of IFREMER became clear. Having tried to minimize the problem, the institute began to defend Doumenge's theories that the alga occurred naturally in the Mediterranean. Henocque must not have understood the significance of the speech that his superiors had obviously dictated to him. Exasperated by his dogmatic recitation, I asked him if he had ever seen the colonization of the tropical alga or if a single member of IFREMER had made eyewitness observations of the algal invasion of our soft-bottom areas that would justify their speaking of them so knowledgeably. His negative response astonished the entire audience. I later regretted having personally offended him. His was only an assessment mandated by a very hierarchically governed organization.

The representative of an environmental protection association of Roquebrune–Cap Martin was amazed at the contrast between the luxurious research equipment (including a flotilla of magnificent oceanographic vessels and midget submarines) used by IFREMER to study abyssal animals of distant oceans and their lack of interest in a problem as important as the algal invasion of our own coasts. A representative of the fishermen asked for the conservation, in a zone closed to construction, of *Posidonia* meadows that were in danger of being destroyed by *Caulerpa*, a suggestion that was squarely in the sphere of interest of the prud'homie. Part of the bay of Roquebrune was coveted by developers of yachting harbors, who had finally abandoned the project because, at the spot planned for the development, underwater, there were meadows of protected species of *Posidonia*.[74] Now, if the *Posidonia* were to disappear, the developers would be able to return! Like many in the audience, I was surprised by this argument, which today can be seen as very perceptive.

The Ministry of the Environment was to have been represented by Lucien Chabason, then a member of the cabinet of Brice Lalonde, but also the candidate of the "Génération Écologie" party from Var in the regional elections that would be held three months later. At the last minute, he could not come because of a completely legitimate personal reason. He sent me a telegram assuring me of financial assistance from his ministry. This was a promise that gave me a bit of hope.

I read the last letter of the representative of the Interministerial

Committee for the Sea.[75] In this way, I informed my audience of his refusal to take charge of the assessment and of his practice of passing me off to a center for the study of algae in Brittany (CEVA). Now, six months after having been given the file on the case by the interministerial committee, the head of the Brittany center had still not seen the alga. It grew, it is true, 1,500 kilometers from his laboratory. I had invited the director of CEVA to the meeting and to the dive. By a telegram received the day of the meeting, I was informed of his desertion. His promise "to examine my copy of the file at the next meeting of the scientific committee" became a joke. Finally, the local representatives of Maritime Affairs, of the Ministry of the Environment, and of IFREMER made an effort to say they were concerned and promised their cooperation.

Well before the meeting, I had solicited an interview with the Prince of Monaco to inform him of the situation and to invite his representative to the meeting. He did not send an official observer; I did not receive his response until the alga spread to the front page of the newspapers. In his response, I recognized the assertions of Doumenge, in whom he had obviously placed all his confidence. In the lecture hall, two staff members of the Oceanographic Museum were friendly and present on their own behalf, rather than as representatives of the Museum; I had greeted them. They made a brief statement admitting that the most likely origin was the Museum aquaria, with the nuance that there were other aquaria in Monaco that could also be incriminated.

More than two years had flown by since my first dive to see the invading *Caulerpa*, at that time found only beneath the windows of the Museum.

Caulerpa taxifolia, Superstar

How the Alga Became a Killer

The December 1991 meeting marked a decisive step in the assessment of the algal problem; the authorities were finally going to have to deal with the invasion. For many people, it had above all constituted the birth of a new media figure—*Caulerpa superstar*. Beginning right after the meeting, many journalists came to my laboratory. In every interview, I was careful to stress the importance of referring to the observations and conclusions of my scientific paper. My colleagues Boudouresque, Cinelli, Pietra, and Verlaque were also questioned. They primarily restated their spontaneous impressions after their first dive to the *Caulerpa* of Cap Martin. They had just discovered the scale of the damage, and their remarks bore the imprint of the astonishing sight of the invaded patches of bottom.

Boudouresque had been the most shocked: "What I saw surpassed my most pessimistic expectations. The prognosis is frightening. A real ecological nightmare."[1] "We are at the threshold of a major ecological catastrophe in the Mediterranean."[2] "This surpasses in extent anything I had imagined. Under the populations of *Caulerpa* there are no other algae, no invertebrates, no molluscs, and practically no fishes."[3]

The inertia of the authorities and our opponents' disinformation campaign certainly contributed to solidify our convictions of alarm, at the same time catalyzing the enthusiasm of several journalists who were firmly convinced by our hypotheses. Some of them were to go further in seeing an imminent catastrophe. Competing with one another, on the lookout for a sharp expression, they exploited any statement that was unduly incisive or had a somewhat misplaced emphasis—sometimes the statements were suggested by the journalists themselves. Most of the time, the articles were dispatched to editors with no further contact with the interviewee. Some transcriptions were very prudent, and others, by contrast, were grossly exaggerated.[4] An examination of the voluminous file of press clippings on this affair nevertheless shows that excesses and errors were, in fact, in the minority. And if there were a few erroneous, shocking, or exaggerated words, it was useless to call the editors, who most often discouraged any request for a retraction or clarification. Nevertheless, it was overwhelmingly the errors that would attract the attention of our detractors.

Here are some of the regrettable interpretations that entered the press dossier on the "marine alien," to use a journalistic expression; it is easy to compare them to our writings (publications and reports) or our recorded broadcasts. Although we had announced that thirty hectares were invaded, people could read: "from Menton to Toulon, the entire Côte d'Azur is invaded." After our statements on the observed attack on the Mediterranean marine flora and the envisioned impact on the fauna, the boldest extrapolation led one to believe that all fishes had fled the invaded zones and that, aside from *Caulerpa*, these zones were entirely sterile. Our hypotheses on the consequences of the toxicity were transformed into certitudes by the press, before being amplified to the suggestion that all seafood would soon be unfit for human consumption.

It was surely foreseeable that unleashing the press would be accompanied by a transmogrification of our declarations in one direction only—that of exaggeration and sensationalism. It was necessary to accept these inevitable inaccuracies in order to achieve the constant goal—to alert the public and thus to make the relevant authorities act.

It was the last chance to make ourselves heard in time. On this level, we had succeeded. This is not the first time that the press has rescued a case from oblivion. At no point did we regret having provoked this media campaign.

The reaction of the authorities is good evidence of the media's persuasive power. While my various attempts to get someone to act had been resolutely ignored for two years, Lucien Chabason, member of the cabinet of Minister of the Environment Brice Lalonde, stated just two days after the appearance of the first articles in the press "that he would work to free up funds for field study."[5]

We had succeeded in getting heard, but there remained the problem of the "minimizers," who were surely going to counterattack. Several reporters tried unsuccessfully to reach the director of the Oceanographic Museum of Monaco. They had to be content with his previous statements.[6] In the absence of testimony from our main opponent, some reporters found items with which to spice up their stories on the affair by turning to IFREMER. Our communication initiatives had evidently forced these experts to reexamine the problem. They increased their declarations of intent to support study of the problem, all the while continuing to appear reassuring. This approach justified their earlier opinions that had led to the inaction of the ministries that had been consulted.

The day after the meeting, Merckelbagh, director of the environment for IFREMER, announced to *Agence France-Presse:* "It's important not to give in to alarmism. It's too early to conclude that we are dealing with a catastrophe. It's necessary to wait for the results of mapping to understand the evolution of the problem and to analyze the human consequences."[7] Two days later, his local subordinate, Henocque, issued press statements that were so ambiguous that they were translated as a plea for *Caulerpa,* leading the editors of *Figaro* to say as a subheadline: "Scientific controversy over an invader."[8] It is true that, until then, IFREMER had shown hardly any scientific interest in this alga. We should recall that, at this time, Henocque and Merckelbagh knew about the invasion of the Côte d'Azur by the alga only from the newspapers or television.

THE PANIC OVER THE TOXINS

There is no doubt that the presence of the toxins was the subject of the most contradictory and excited interpretations. Some journalists thought that we were too cautious about the risks. We were bombarded with questions of nuance: was it going to poison all the fish? Are the fish going to die from it? Our negative answers disappointed those who loved scoops. Little by little, a succession of erroneous interpretations or exaggerations tended to support the idea that it was not only the environment that was threatened, but humans. Pell-mell, the consumption of seafood, fishing, seashore tourism could all—would all—be affected in the short term!

The prize for catastrophism can be awarded to a reporter from Var for his description of a diver emerging from the water with a clump of *Caulerpa taxifolia:* "The deceitful, fine-toothed leaves lose their sparkle. They resemble millipedes mixed together in a disgusting mass." "The alga threatens to make the consumption of all seafood from the coast dangerous!"[9]

On land, we come in contact every day with plants that, if ingested, could be toxic. In the sea, plants that can be termed "toxic" are less common. However, certain algae can contain or secrete substances that serve as repellents and thus protect the plant against herbivorous animals (fishes, sea urchins, or molluscs). Some marine animal species have adapted to these toxins. Some algal toxins can even accumulate in the flesh of herbivores, rendering them unfit for human consumption (this phenomenon of accumulation is rare in terrestrial animals; if, by accident, they ingest toxic plants, they usually become sick, just as humans do).

In tropical seas, the ciguatera, nicknamed the "itch," is a real scourge that prevents the exploitation of the natural marine bounty of many atolls and isolated islands. A microscopic alga comprising a single cell, *Gambierodiscus toxicus,* a benthic dinoflagellate, first identified in the Gambier Islands of Polynesia, contains toxins (maïtotoxin, ciguatoxin, and scaritoxin). This alga is eaten by some fishes and crustaceans, which are in turn eaten by humans. These people feel an itch (the "itch"), then suffer gastric problems or muscle pains. The gravest symptoms are reversible paralyses that persist for various lengths of time. Magnificent fish, crustaceans, or molluscs can be edible on the eastern side of an atoll and highly pathogenic on the western side; it all depends on the presence or absence of the unobtrusive microscopic alga. Only the natives know the subtleties of such cases.

In Japan, one fish species costs almost as much as caviar: the tetrodon or fugu. Its gustatory qualities are highly prized by the Japanese. But Asian cooks have to be officially licensed to prepare it, because lethal toxins build up in its liver and blood. Despite all the precautions, several lamentable deaths occur each year following ingestion of this fish. The lethal agent of this "Japanese roulette" is a toxin produced by a bacterium. The bacterium is eaten by microorganisms that, in turn, are eaten by the fugu. The microorganisms and the fugu are immune to the toxins, but humans are not. The fugu was thus protected from overfishing by humans until the Japanese found a way to eat it without perishing. In Europe, we have all heard about periodic prohibitions against selling molluscan shellfish. In most cases, unpredictable eruptions of microscopic algae are responsible. The shellfish filter the water, ingest the toxic microalgae, and become toxic themselves. The shellfish are healthy, but not the humans who eat them. To avoid epidemics of gastroenteritis, prolonged diarrhea, and neurotoxic syndromes, the sale of shellfish is banned while we wait for the responsible microorganisms to disperse spontaneously from shellfish beds.

Toxicity in the genus *Caulerpa* has long been known, and, as I have noted above, I had every reason to think that the species introduced to the Mediterranean contained the same toxins. Certainly, the effects of this toxicity remained anecdotal in the tropical oceans, where *Caulerpa*

74 species are far from dominant in the coastal flora (tropical fish have many other algae to nibble). But the situation is very different in the sites invaded on the Côte d'Azur, where the alga tends to eliminate the majority of the local flora and to become the dominant vegetation. It was thus legitimate to ask about the risks incurred by letting such an invasion run its course: would the herbivorous animals still find enough food? Would some species eat *Caulerpa* exclusively and thus accumulate toxins in their flesh?

When these questions were raised in our scientific paper, they were not criticized by the three anonymous referees. These issues were also raised in May 1991, in the opinion issued by IFREMER, and in August 1991 in the national press by the spokesperson of that institute.[10] The reasoning was never questioned in subsequent works. But it is true that the toxicity of a marine species, an alga, seems strange and worrisome. To respond to the curiosity of reporters, we had to distinguish certainties from hypotheses. Despite all my efforts on this delicate matter, I was unable to avoid regrettable interpretations.

The most deplorable aspect of this affair was the surprising absence of any dialogue between the researchers who reported the available information on the subject and the advisors of the national authorities. These latter experts had not even bothered to verify the presence (or absence) of toxins in the alga, as we had recommended as early as July 1990. This is why we sought out specialists on our own. These specialists presented their conclusions at the meeting in Nice in December 1991, confirming my hypotheses in the *Oceanologica Acta* article. The verified presence of toxins supported the inference of an aggravated ecological impact; it was logical that a vegetation dominated by *Caulerpa taxifolia* would not be appreciated by the "vegetarian" fauna. It remained to be seen if the growth of masses of toxic algae on marine soft bottoms was capable of leading to feeding adaptations and to the accumulation of toxins in a new food chain. A representative of the National Institute of Health and Medical Research (INSERM) was eager to undertake this ecotoxicological research.

As fishing was under its jurisdiction, IFREMER was going to focus its attention on the toxicity theme, in a way that was quite surprising. **75** The day after the meeting at Nice, where the Italian researchers' studies proved that toxins were well represented in Mediterranean *Caulerpa taxifolia,* Merckelbagh, trained as an economist and now administrative director of the environment at IFREMER, communicated his view from his Paris office: "The evidence on this toxicity is not convincing."[11] He thus specifically contradicted the previous opinion of his institute. In May 1991, IFREMER had stated, "The toxins (caulerpenyne and caulerpicine) play an anti-herbivore and antibiotic role. They are suspected of having a toxic effect on humans through the food chain." IFREMER even stated specifically that "effects similar to ciguatera have been reported, but not mortality." The sudden skepticism expressed by the director of the environment was not based on any new research or logic.

Three days later, Henocque was busy echoing his superior's about-face. A reporter from *Figaro* asked him if *Caulerpa* was dangerous. He replied: "Nothing has been proven for the Mediterranean, or even in the tropics, to my knowledge."[12] However, Henocque was a scientist. He knew how to read the publications on the subject, and he had heard the presentations by Pietra and Boudouresque. In the same article, the reporter contrasted this statement with my presentation of the problem: "According to Alexandre Meinesz, the alga produces toxins to defend itself against grazing fish. An Italian laboratory has isolated six toxins, which are in the process of being analyzed and tested. Because, if it is toxic, that magnifies the problem. But, more than the toxicity, it is its very presence and explosive growth that worry Professor Meinesz." Nor did Doumenge hesitate to go still further in entering the media debate: "The toxicity of this alga has not been 100% proven."[13]

We were faced once again with claims that lacked a supporting scientific basis. In fact, the declarations by IFREMER rested on an interpretation of the word "toxic." For their spokesmen, it seemed as if the toxins would have to affect humans in order to be considered toxic. And as no impact on humans had yet been observed, the alga could not be called toxic. Using this nuanced rhetorical device, IFREMER representatives chose to challenge the hypotheses of "alarmist" scientists on tox-

76 icity rather than to correct the major errors on toxic impact that turned up here and there in several newspapers.

To limit the effects of the ill-will engendered by the question of algal toxicity, the scientists spoke out. Boudouresque, without denying the presence of toxins in *Caulerpa taxifolia*, emphasized the fact that animal species were unable to feed on this *Caulerpa*. He explained, correctly, that, as the toxins were what rendered the alga inedible to animals, this fact fortunately eliminated the risk that the toxins would get into the food chain. For my part, I focused my remarks on the problems of ecological equilibrium. However, after the December meeting, I had ceased to comment on toxicity, advising my questioners to ask the INSERM specialists who knew more than I did in this area and who could explain the tests that they were going to run in order to know more. During the months that followed, two researchers from INSERM told the media, without passion or exaggeration, the principal outcome of their investigation of the toxins.

The main concern of IFREMER was to reassure the public, even if it meant denying the existence of the toxins. This attitude doubtless had the effect of reinforcing the curiosity of the press. Intrigued by the observation of a major discordance, and having before them the many documents we had provided, many reporters wanted to understand the attitude of the government advisors on matters of fishing and the marine environment. We were witnessing a veritable dialogue of the deaf, which could be summed up as follows:

The scientists: The alga is toxic, and we must study the consequences of the toxicity of a vegetation that is spreading and tends to dominate other plants.

Doumenge and his associates: The presence of the toxins is unproven; it's all a tall tale designed to tap research funds.

The experts of IFREMER: There are no toxins.

Some reporters: Where there's smoke there's fire. It's obvious: the toxins adduced by the scientists, and denied by the director of the institute

suspected of having introduced the alga and by the government experts, are going to contaminate seafood.

AND *CAULERPA* BECOMES A KILLER

In December 1991, *Caulerpa taxifolia* was christened a "killer." This terrifying descriptor first appeared December 5, in an article in *Var-Matin*,
signed by Roland Tardy. It surfaced in the sub-headline of an interview
with Boudouresque: "The 'killer algae' (*Caulerpa taxifolia*) are literally
invading the Mediterranean coasts: danger! Warning by Professor C.-F.
Boudouresque. . . ." In *Nice-Matin* of December 17, the reporter Philippe Fiammetti used the adjective both on the front page and in the
headline of the report on the meeting held in Nice the day before: "The
ecological nightmare of the killer alga"; "We must rip out the killer
alga."

I was hardly surprised by this journalistic brainstorm, which led
to other strong qualifiers, such as the "tentacled alga," the "hallucinogenic alga," the "toxic alga," the "crabgrass of the sea," the "alga that
is smothering the Mediterranean," the "green scourge," the "marine
monster," the "sea poison," the "alga that swallowed the Mediterranean," the "formidable stranger," the "green cancer," the "green plague,"
"AIDS of the sea," etc. The alga touched a nerve, and every reporter
sought a catchy nickname. Doubtless, they had only wanted to focus
readers' attention on the fact that the alga *killed* the marine flora of the
Mediterranean, dominating the other plant species while competing for
space or light. But the expression "killer" struck a subconscious chord,
and curiosity took reporters and the public much further.[14] Also, was not
this all happening in the South of France, a region where the use of
metaphor to illustrate anything that appears out of the ordinary is completely normal?[15]

"Killer alga" quickly became the preferred nickname in the press,
in France as elsewhere (killer alga, green killer, alga assasina, alga ases-
ina). Appearing first during the cacophony over the toxicity of the alga,

the expression was bound to unleash simultaneously anxiety in tourism circles and disapproval among senior bureaucrats and local councillors. Many leading Paris bureaucrats even suspected me of creating the adjective and promoting its use! Thus named, the alga aroused fear and moreover conferred a negative image on the Côte d'Azur. According to several local politicians, many tourists telephoned mayors' offices or tourist bureaus to be assured about the total absence of risk associated with swimming in places where the killer alga was growing. Local elected officials increased the number of reassuring statements and fingered the university alarmists as troublemakers. Fearing that the "beast" was going to make the bathers flee, some politicians, no doubt afflicted by the "Jaws" syndrome, denied the existence of the alga on their coasts or asserted that it existed only in the imagination of eco-scientist Cassandras. Diving clubs and committees for tourism promotion bragged in the press about the quality of the waters of their "*Caulerpa taxifolia*-free"[16] shores.

Once launched, the nickname was bound to spread. Even if I had never used it in interviews, even if I had said that I was ready to swim with my children above *Caulerpa* prairies and to eat any local fish, it would not have made any difference. I even tried to make the journalists act responsibly, pointing out to them that the expression cast a bad light on all algae, a group that has not, it is true, always had a great reputation (unfounded fears, as algae are neither stinging nor barbed). They replied that, even if I was simply trying to do a good job as a scientist, they too were trying to do their jobs as reporters well; that in any case it was too late because the alga was already known by that name; and finally, that this "slight" exaggeration had increased newspaper sales.

For his part, Boudouresque routinely began his interviews by pleading with reporters not to use the adjective "killer." Wasted effort! A local newspaper published an article headlined, "Who said 'killer alga'?" — in which one could read the appeal by Boudouresque, as depicted by the reporter, as pathetic: "I'm begging you, don't talk about the killer alga any more. This term doesn't correspond to reality!"[17] There was no better way to make the nickname stick. And when some understanding correspondents promised to avoid using the term, the editors went over their heads.

This sobriquet conferred a distinct prejudice on all those who fought the alga. Whereas we were already called purveyors of catastrophe, the overwrought metaphor was ascribed to us by our detractors to show that all our remarks were exaggerated, and even fantastic! How many times had we heard our opponents say that the alga had never killed anyone, that those who hawked such rumors were professional alarmists, having blown up the affair in order to increase the research funding of our laboratories! This new polemic had, as its main effect, given a reprieve to the alga. People were busy trying to eliminate the controversial word from the media rather than the alga from the sea.

By dint of being overused, the adjective "killer" lost its impact, to the point of becoming a pseudonym of *Caulerpa taxifolia.* From that point on, it was futile to struggle against the use of this frightening term. Despite the regrettable excess, our objective was achieved—it was in fact our sensitization campaign that triggered the whole brouhaha. The final attempts of the "minimizers" to sow doubt on the existence of toxins or their importance were destined to be swept away by a whole series of unanticipated sudden developments. After a period of fruitless media debates, the "affair" literally exploded.[18] The public and the decision-makers were overwhelmed by talk of an irreversible threat to marine ecosystems. We had precipitated a crisis, and we knew it was going to produce a reaction.

Late December 1991: The Fishermen's Representative takes up the Baton

The day after the meeting at Nice, I received a call from Augustin Aquila, president of the regional committee on fishing and mariculture of the Provence–Alpes–Côte d'Azur region. Alerted by the press, he reproached me for not having invited the fishermen to this meeting. In fact, I had invited the prud'hommes[19] of the eastern part of Alpes-

80   Maritimes and the chairman of the local committee of fishermen in that department. But only one representative of the fishermen showed up.

Aquila wanted to raise the consciousness of the fishermen of the French Mediterranean coast to the problem posed by the introduction of the alga. He suggested organizing a meeting for this purpose the following week in Toulon. Inviting my participation, he asked me for all the available scientific documentation. He invited all the prud'hommes of the Côte d'Azur, Provence, and Languedoc-Roussillon, but also all the local and national authorities. The invitation was alarmist: "The gravity of the consequences of the proliferation of the alga *Caulerpa taxifolia* in marine habitats with respect to the economics of fishing, as well as to ecology and tourism, leads me, before this phenomenon becomes irreversible, to ask you or your representative to attend the meeting, the urgency of which will be obvious to you."

After the recent press coverage of the entire affair, this executive of a well-respected organization had no trouble attracting an audience among those who were unable to attend the Nice meeting, or who had not deigned to do so. The meeting was held just before Christmas, December 23, 1991, in the grand lecture hall of Maritime Affairs at Toulon. A hundred people attended the discussions: regional councilors, departmental councilors, representatives of Maritime Affairs, IFREMER, prefectural services, representatives of fishermen, yachtsmen, and divers, not to mention the many reporters that Aquila had invited.

In turn, Boudouresque, Verlaque, and I presented the facts we had already discussed at the Nice meeting. Two representatives of IFREMER described the position of their institute. Coming from many prud'homies, the fishermen were worried. They asked myriad questions about what caused the introduction of the alga and about the apparent lack of response to the phenomenon. On this matter, IFREMER showed its colors once again through the speech of an ichthyologist from their station at Sète, who defended the Suez Canal origin hypothesis advanced by Doumenge.[20] At the end of the meeting, Aquila demanded, on behalf of the fishermen, that all possible measures be taken to battle the alga and that an administrative inquest be undertaken to determine who was responsible for introducing it. A motion reflecting these demands was sent to all relevant authorities and included requests for more information on a means of stopping the invasion, necessary funding to do so, and increased community involvement in eradication efforts.

In the meantime, the alga had been found in a yachting harbor very far from Monaco, at Saint-Cyprien on the Catalan coast, near the Spanish border. There were only about five square meters of *Caulerpa taxifolia* growing on the bottom of the harbor. A diver responsible for maintaining the harbor had recognized it thanks to the pictures that had been circulated by the media after the Nice meeting. He eradicated the colony by hand and gave samples to the Arago Oceanographic Center at Banyuls. Questioned by the local press, a phycologist from the center called on divers to explore the whole area and to report the presence of the alga.[21] Several days later the harbormaster released an announcement that divers had rigorously scoured the waters of the harbor and no trace of *Caulerpa* had been detected after the eradication.[22] A member of the Sète branch of IFREMER collected several fragments of the alga, the first samples of *Caulerpa taxifolia* actually examined by IFREMER since the beginning of the affair. He presented them at the Toulon meeting and announced this distant range extension to the assembled reporters.[23]

This site was almost 350 kilometers in a straight line from Monaco. The origin of the alga and its dispersal modes were discussed: "eggs" (zygotes) drifting in the dominant currents? Fragments carried by the anchor of a boat that had stopped at Cap Martin? Purging of a local aquarium? Since it was known that the alga grew profusely in the vicinity of Monaco, many aquarists had come there to gather hundreds of kilograms of *Caulerpa taxifolia* to sell. Might it subsequently have been thrown out in the Saint-Cyprien harbor? For my part, I had a hard time imagining an aquarium enthusiast who went to the trouble of carrying an aquarium, or its contents, in order to dump them on the bottom of a harbor where conditions for aquarium flora and fauna are very suboptimal. Nevertheless, after the suggestion of this hypothesis, the Sète fishermen took it upon themselves to convince the directors of aquaria to eradicate any *Caulerpa* in their tanks. This was how the thalli of *Caulerpa taxifolia* that were displayed in all the public aquaria on the Languedoc-Roussillon coast came to be killed.

The fishermen's meeting and the announcement of the discovery of the alga in Pyrénées-Orientales made a big splash. *Caulerpa* superstar, alias "the killer alga," became a media prop and was exploited as such.

January 1992: The "Ambush" of Saint-Cyr-les-Lecques

On January 13, 1992, we received in our laboratory a phone call from the president of a dive club, the Aquanaut Club, based at the harbor of Saint-Cyr-les-Lecques, a site on the western Var coast—*Caulerpa taxifolia* had been found the day before in the harbor. By chance during a dive, a club member had discovered one square meter of the alga. Aware of the problem, he surfaced, got a bag, dove again, tore out all the algae and brought them to the surface. The club director gave me an accurate description of the alga, which he had preserved in a cask. No doubt was possible: it was certainly *Caulerpa taxifolia.*

Wanting to get rid of the alga, he sought my advice. The approach of tossing it in the household garbage to be incinerated was the most logical. However, I wanted to examine the alga first to make several observations on the size and shape of the fronds. I also asked if he would agree that the local reporters (from *Var-Matin* and *Le Provençal*) should attend my assessment so that the divers of the western part of the department of Var and the nearby department of Bouches-du-Rhône would be inspired to explore neighboring patches of bottom. Arriving at the harbor on January 20, 1992, we were surprised to find a small crowd in front of the dive club headquarters, including five or six reporters, Henocque of IFREMER, Verlaque of CNRS in Marseilles, and divers from La Ciotat. A city councillor gently reproached me for not having warned the mayor that I was coming. What had attracted so many people? In fact, I had underestimated the notoriety of the alga.

The president of the Aquanaut Club had not come, and we were
greeted by the diver who had discovered the alga. He partly emptied
the cask of *Caulerpa* onto the ground. He drew my attention several
times to rock fragments as large as a fist covered with concretions and
with *Caulerpa*, rocks of a type not normally found in the harbor, according to him. Questioned from all sides, I was not especially interested
in these rocks. I thought that they were simple bioconcretions[24] similar
to those that fishermen often gather in their nets and subsequently toss
into harbors. I was much more intrigued by the remarks of divers who
came from the neighboring town of La Ciotat, where people had just
reported to them that the alga was present near the Ile Verte, opposite
the harbor of La Ciotat. Between the inspection of an area from which
the alga had been eradicated and research on a possible new colony, I
had no difficulty making a choice. After presenting the "certificate of
scientific cooperation"[25] to the diver and making several statements to
reporters, I followed the divers from La Ciotat and let Verlaque take the
algal samples as well as the strange rock chunks. At La Ciotat, we dove
in vain searching for the alga; either the information on the site was too
imprecise or the identification had been mistaken.

The next day, the local press heralded the discovery of the alga at
Saint-Cyr-les-Lecques, emphasizing (as I had hoped) the importance of
its having been reported and systematically ripped out.[26] Questioned,
the mayor of Saint-Cyr-les-Lecques, Jean-Pierre Giran, declared himself
"vigilant and preoccupied" by the appearance of the alga in his town.
He took the initiative of sending municipal firemen to inspect the area
around the harbor.[27] In the January 21, 1992 edition of *Var-Matin*, the
presence of the alga occupied the majority of the first page, with the
headline reeking of catastrophe: "Saint-Cyr: Killer Alga Alert!" This
news item was treated as of equal importance with the rest of the day's
happenings, which were not trivial: an A-320 Airbus had just crashed
into Mont Saint-Odile, near Strasbourg.

This incident could have ended with these news stories. But three
days later, I learned from the national press about the sudden fury of
the mayor of Saint-Cyr-les-Lecques. Not only was he irritated by not

having been informed of my visit to his town, but he found my actions suspect. He criticized me for having announced the presence of *Caulerpa taxifolia* and for having invited reporters even before I had seen the alga. His suspicion was motivated above all by a revelation: the chunks of rock, examined by a researcher in Boudouresque's laboratory, were in fact old coral debris from the Red Sea or Indian Ocean. The researcher, who informed the mayor of this fact, naively suggested to him that he not disclose it, which to the mayor seemed even more suspicious. Suspecting us of deliberately mounting a media production, he lodged an official legal complaint against parties unknown.

The press got hold of the story, drawing hasty conclusions: "Killer alga: an ecological crime! Formal complaint lodged. Killer alga: the trail of an ecological crime."[28] "Killer alga formal complaint."[29] "Killer alga: suspicions of ecological crime."[30] "Police enter the affair of the algal threat."[31] On the front page of *Var-Matin* was a photo of the mayor surrounded by firemen and the caption: "Toxic alga: the stone in the sea. Saint-Cyr: deliberate pollution scare. According to the mayor of the town, J.-P. Giran, the presence of the toxic alga in the port of Lecques is due to a malevolent act."[32] The article itself was equally incisive. With the sub-headline, "Saint-Cyr: deliberate pollution scare," the "crime" was described thus: "Criminal spreader of the alga. This is a serious matter. If the inquest undertaken following the lodging of a formal complaint against parties unknown shows the claims are well founded, the entire file on the toxic alga will have to be examined: Are we witnessing a veritable plot hatched for as yet unknown reasons, or rather is someone trying to aggravate an already alarming situation? Who can guarantee today that some simpleton hasn't deliberately sown clumps of *Caulerpa taxifolia* along the coast?"

The thesis of a willful biotic pollution of the Mediterranean coast was thus launched. The local correspondent of *Var-Matin* had access to the front page and several columns of the national newspaper *France-Soir* with the following analysis of the matter: "Today a frightful question is being asked: has someone been trying, from the beginning of the epidemic, to contaminate the entire Mediterranean with this species of

marine cancer? If such an ecological crime has truly been committed at
Saint-Cyr, one can ask if earlier appearances of the toxic alga ... were
not the products of a warped scheme."

I had a hard time deflating this veritable collective paranoia. I tried
to turn the inquiry in the direction of a thoughtless act by an aquarist.
The press took up my statements in which I emphasized the growth in
commercial trade in the alga that came about because it was easy to
harvest at Cap Martin. I also recalled for them my initial analysis of the
problem, in which I advocated the regulation of introduced species in
aquaria. I showed my 1991 letter to the Prefect of Alpes-Maritimes, asking for the urgent prohibition of commerce in the alga. Philippe Fiammetti, reporter for *Nice-Matin*, investigated aquarium stores and found
that it was for sale everywhere.[33] To his question, "Do you have any
plants for my tropical aquarium?" a shopkeeper answered, "We still
have some very beautiful *Caulerpa* left, but just a little. We expect to
receive a big shipment in fifteen days. This is an alga that survives very
well, and, with good light, it grows without problems."

I telephoned the mayor of Saint-Cyr-les-Lecques, begging his for-
giveness for having unintentionally offended him by not alerting him
to my arrival. I swore to my good faith and complete innocence in the
appearance of the alga in his town. In fact, I had not realized how much
the media stardom of the alga had turned every new appearance of it
into an official and attractive event. During the previous four months,
the announcement of my discoveries of the alga at Lavandou and Agay
had excited little interest; I had never been invited by the local authorities to attend the official validation of the discovery of an algal colony. I
had a lot of difficulty reestablishing good relations with the mayor. In
February, a meeting of mayors of the Côte d'Azur was organized at Menton to consider their position on the algal invasion problem. Mr. Giran
did not participate in this meeting, but he sent a letter detailing his
suspicions of me.

During the same period, his candidacy in the regional elections was
announced. Elected in March, two months later he was appointed to
the presidency of the Commission on the Environment of the Regional

Council of Provence–Alpes–Côte d'Azur. For a year I had been president of the Environmental Section of the Social and Economic Commission of the same region.[34] This role implied the need for good relations with my regional elected counterpart as we examined the same files. I heard rumors that our bad relationship could force my resignation. In June 1992, I went to his office to clarify my position. Despite all my explanations, I failed to convince him. Although the intense media coverage of the Lecques affair had subsided, he still wanted me to confirm at every opportunity the absence of the alga in his town. In September 1992, local divers found thalli of *Caulerpa taxifolia* in the same spot. This was probably a matter of snippets of the plant that had been missed in January. I was immediately notified. The divers assured me that they had meticulously inspected the site and eradicated the *Caulerpa*. Since my main hypothesis was that this reappearance was both logically expected and anecdotal, I did not return to Saint-Cyr-les-Lecques and the press was kept out of it. Ironically, one year later, the mayor of Saint-Cyr-les-Lecques took up the torch in the struggle against *Caulerpa*. He promoted the efforts of two of his citizens, amateur aquarists, who had discovered an algal eradication method employing an electrolysis system with copper ions.

As for the diver who had found the alga in the harbor of Saint-Cyr-les-Lecques and had surfaced with the chunks of coral, I was surprised to see him in uniform ten days after his discovery aboard a vessel of the French Navy (he was an employee of the Office of Maritime Affairs of the Maritime Prefecture of Toulon). He had just participated in a military operation to remove *Caulerpa taxifolia* at Méjean cove near Toulon, requested by the Secretary of State for the Sea, who had wanted to see a sample during his visit to Toulon.

Information provided by the mayor showed that the corals found with *Caulerpa* were not located at the base of the wharf. According to him, it was impossible that they had simply been tossed from the wharf; they had been placed by a diver at buttresses of the wharf, as if someone had taken care to put them in an obvious place or to avoid their getting stuck in the harbor mud. Even though the diver, the only person to have

seen the exact configuration of the coral on the bottom, had a more   **87**
qualified opinion on this hypothesis and thought that the corals could
nevertheless have been simply tossed into the sea, he admitted that the
arrangement of the pile of *Caulerpa* and coral was very strange. As in
the case of the *Caulerpa* found in the Saint-Cyprien harbor, it was
difficult for me to envision an aquarist emptying an aquarium off the
end of the wharf. At that time, fragments of *Caulerpa* had some value.
And aquarists with tropical aquaria know that animals and plants,
bought at premium prices, can be repurchased by an aquarium shopkeeper (of which there is one in Saint-Cyr-les-Lecques).

Involved in the affair by the formal complaint, the police conducted
a sustained inquest for over a year. I was never questioned. This affair
was exploited to different ends by various persons. Despite the suspicion
displayed towards me by the mayor of Saint-Cyr-les-Lecques, I now
think that I was never his main target. He had certainly wanted to be
seen during this electoral period as concerned and active in the face of
"the killer alga proliferating in the waters of his town." His constituents,
who made their living from seafood and tourism, had suffered from the
presence of the alga (fish sales slumped immediately following the discovery of the alga at Saint-Cyr-les-Lecques). They surely approved of
the steps taken by their elected official.

The directorate of the Oceanographic Museum of Monaco exploited
this confused affair, heralding it as proof that the alga could have come
from any aquarium. Although no journalist had been able to reach him
since our meeting at Nice, Doumenge ended his silence to stress to the
press the innocence of the Museum with respect to the origin of the
contaminant: "In police thrillers, the path that seems the most obvious
is not always the best."[35] He again stated his favorite hypotheses: the
arrival of the species through the Suez Canal or the sudden flourishing
of a dormant species because of a climatic change such as global warming. He added that he was going to prove scientifically that the alga
strengthened environmental equilibrium by producing oxygen. One of
his friends later recalled the confused hypotheses about the introduction
of the alga to Saint-Cyr-les-Lecques, which tended to exonerate the

88 Museum.[36] My hypothesis of a single origin for the invasion was thus questioned. A thoughtless isolated act by an aquarist or an "ecological crime"—both of these potential causes tended to attenuate the responsibility of the Museum. By the same token, the justification for an eradication campaign against an alga, deposited in one place and spreading from there, would have to be reconsidered if it could be shown that the invasion was due to multiple, uncontrollable acts.

The paranoia over the "ecological crime" also disturbed me. I considered the possible perpetrators of this diversion, who had surely foreseen neither the intense local exploitation of this affair nor the police inquest. I retain from this confused affair the impression of having fallen into an ambush. One thing is certain—the little square meter of *Caulerpa taxifolia* at Saint-Cyr-les-Lecques resuscitated the notoriety of the "killer alga." Some profited from this turn of events, others came to regret it.

The National Government Takes Control of the Affair

Local scientists who coordinated research and took it upon themselves to assess the problem and to organize a press conference; fishermen who spread word of the risks; a central administration that is suspected of negligence; a killer alga that produced a general paranoia—it was all happening in the South of France, a place where exaggeration of everything is suspected. This situation had deeply upset the Secretariat of State for the Sea and some senior officials of IFREMER. It was necessary for them to regain control of the affair, which was getting out of hand.

Forced to address the problem, the technocrats of our national agencies were unconvinced by the scientists' arguments. They saw one urgent task—to stop the media drumbeat about a catastrophe. Their strategy was to create well-supervised committees so that the decisions could be controlled and their consequences made predictable. The operation was undertaken briskly. I had a sense of foreboding about the threat of

a government takeover. At the meeting at Nice, I had formed a scientific
committee uniting the affected researchers, with the goal of impressing
upon the authorities the major risk posed to Mediterranean ecosystems
by the proliferation of *Caulerpa taxifolia*. This self-proclaimed committee, limited to four French and two Italian scientists,[37] took as its objective the coordination of research and the communication of its results
to the relevant authorities. We had hoped for the participation of an
IFREMER expert, if only to show our good will towards and scientific
cooperation with those who, until then, had neglected and deprecated
the problem. The fishermen's representative had also agreed to join the
committee as observer.[38] The press hailed this concrete initiative: all the
experts had to be united.[39]

In order to maintain the momentum given by the sudden and wide
media attention to the affair and to unify our new allies better (users of
the sea—fishermen, divers, and yachtsman—who had flocked to Toulon), I suggested to the regional fishermen's representative that a battle
committee be created to encompass all interested parties. Fearing that
IFREMER would co-opt such a committee, I stressed the utility of its
independence relative to both the scientists and the authorities. It ought
to be a pressure group that would allow its members to gather all the
facts on the evolution of the phenomenon and to demand action in keeping with the gravity of the situation.

On Christmas day, I was finally able to relax; I had truly done all
that I could have so that the problem would be known and dealt with.
I thought I had taken enough precautions that the affair would evolve
in the right direction, that the alga would finally be taken care of, and
as quickly as possible. Unfortunately, I was mistaken yet again.

The last day of the year, a reporter faxed me a press release distributed the day before Christmas by IFREMER. It showed me how the
affair was going to be treated. Entitled "*Caulerpa* and the researchers,"
the release concluded: "To specify the extent and the potential ecological consequences of this phenomenon of proliferation, IFREMER has
decided to engage in collaboration with university laboratories, CNRS,
and INSERM in a process to define the research agenda, both basic and

applied, that could be supported by the public. Very quickly, a coordinating committee will be established to direct these studies and will include administrators as well as government marine experts and scientists."[40]

I was thus able to see that IFREMER, while announcing its establishment of this process, cited the very entities participating in the scientific committee that we had established, but without explicitly naming the committee. But, above all, IFREMER announced exactly what I had feared—the creation of a coordinating committee including scientists, administrators, professionals in the social sciences, and elected officials, a structure that could only ratify the reassuring opinions of IFREMER. This is a classic strategy tested in many other circumstances; it lowers the stakes, reduces the level of excitement, and allows the national authorities to control the situation. At the end of the day, it is nonfunctional.

The regional fishermen's representative had also received the IFREMER press release, which he hastened to publish in his monthly bulletin.[41] I was not able to convince him of the utility of creating an independent entity comprising divers, fishermen, yachtsmen. He instead sent to the three prefects of the Mediterranean regions[42] (Corsica, Provence–Alps–Côte d'Azur, and Languedoc-Roussillon) a request to create a coordinating committee such as IFREMER had announced, offering control of all the pressure groups to the national government.

A week later, a special meeting of the senior bureaucrats of the agencies concerned with *Caulerpa* took place at the Paris headquarters of the Secretariat of State for the Sea, under the aegis of the Interministerial Committee for the Sea. I was to be the target of harsh criticism. The scientists and government experts who were setting out the situation were all IFREMER members. To keep me away, Boudouresque was asked to constitute an official scientific committee, thereby invalidating the one that we had created several weeks earlier. Boudouresque set only one condition—that I be his co-chair. I accepted this "placement under supervision"; my position as symbolic ringleader would not allow me to direct an official scientific committee calmly. My goal was not to shine personally but to generate the strongest possible struggle against the alga. The main objective that I had focused on for two years had been achieved—the affair was no longer being handled with indifference. I hoped my "surrender" and my integration into the new structure would contribute to defusing the tension. Aside from normal dialogues on the choice of meeting dates, I received no letter relating to my official title; all communications were via Boudouresque.

Under Boudouresque's direction and with the help of the scientific laboratories (as had been shown since the beginning of the affair), a research program was elaborated during the first two weeks of January 1992. It would cost about two million francs and would engage several institutions (INSERM, CNRS, IFREMER, and three universities). Boudouresque, whose research abilities were matched by his talents as an organizer and mediator, possessed a remarkable capacity for work. He applied all his energy to taking up and fleshing out my modest research proposals,[43] presented in vain to the ministries ten months earlier.

CREATION OF THE OFFICIAL SCIENTIFIC COMMITTEE

The new committee, named the Scientific and Technical Committee, was created after consultation between IFREMER and Boudouresque. The "technicians" in fact had more representation than the "scientists"; the four French scientists, founders of the first committee, were surrounded by four military personnel representing the maritime prefecture and the local and regional offices of Maritime Affairs, four members of IFREMER, six representatives of various organizations— Ministry of Research and Technology, Interministerial Committee for the Sea, fishermen, aquarists, the maritime service of the department of Bouches-du-Rhône (where no *Caulerpa* had yet been reported)— and two external scientists (a toxicologist from Nantes proposed by IFREMER and a Dutch phycologist proposed by Boudouresque).[44]

This committee met for the first time on January 21, 1992, at Boudouresque's headquarters at the University of Marseilles-Luminy, convened officially by Jean-Claude Hennequin, interregional director of

92 Maritime Affairs for the Mediterranean. The scientific program, prepared by Boudouresque, was presented after approval by IFREMER.
IFREMER had been charged with studying the feasibility of eradication. On this delicate issue, I dispassionately drew the attention of my
colleagues to my conclusions, convinced it was more a matter of time
than a technical matter. The report in which I had proposed a battle
plan (presented to the Prefect of Alpes-Maritimes at the end of November 1991) was made known to the committee. The subject was vexing;
the majority of committee members wanted to know more about this
alga that they were now having to learn about. Most members had never
seen it. Greater importance was thus placed on the general assessment
of the problem. Management actions were nevertheless discussed, such
as the eradication of several dozen square meters of *Caulerpa* along the
Var coast. But no date and no financing was specified for management.

I was disappointed by the legalistic approach of the assessment; it
took no account of the dynamic aspect of the problem, of the alga that
continued to thrive in our waters. In six months, the alga was going to
spread and be disseminated anew, and the situation would become less
and less controllable. I was torn between two sentiments—on the one
hand, waiting patiently for confirmation of my observations and deductions, and on the other hand, fear that inaction would lead to free rein
for the alga. It was the second sentiment that I always expressed; at
every committee meeting, I reminded a skeptical and annoyed audience
of its responsibility. The more time passed, the greater the cost of eradication, no matter what technique was used. Beyond some threshold,
which I specified as the summer of 1992, eradication would become definitively impossible. My objections were unheeded, and I quickly understood that the problem was going to founder in a textbook procedure,
well known in attempts to resolve controversies. The passage of time
would inevitably cause passions to wane. The alga would be able to
spread and to threaten everything I loved to see underwater—the biodiversity of the Mediterranean seafloor.

Boudouresque fulfilled his mission, to coordinate the scientific assessment of all facets of the problem. In fact, our roles were complemen-

tary, as he was convinced by my analysis of the situation. It was useful   **93**
to demonstrate, according to the rules, that the alga did, in fact, constitute a grave ecological problem. I was useful by virtue of my alarmist
interjections, which he could no longer hope to express. By taking positions always interpreted as unduly pessimistic or out of order, I gradually became the scapegoat in the eyes of the ministerial technocrats,
while the other co-chair, who became more reserved and more conciliatory, was considered as more reliable.

At the end of the first scientific committee meeting, Boudouresque
organized a press conference to announce and to explain the research
program that had been adopted. The media publicized the information
at the exact time when the Saint-Cyr-les-Lecques affair was becoming a
sensation; the paranoia regarding the "ecological crime" to some extent
masked the presentation of the projected research.

ESTABLISHMENT OF THE COORDINATING COMMITTEE

The Coordinating Committee announced by IFREMER the day before
Christmas was constituted in parallel with the Scientific and Technical
Committee. It was to engage, satisfy, and involve local elected officials
and representatives of pressure groups (fishermen, divers, yachtsmen).
The latter were broadly integrated into the effort. The Coordinating
Committee was supposed to gather scientific information from the scientists in order to decide which actions to undertake, and to coordinate
these actions. To attract public support, its establishment had to be solemn, well publicized, and reassuring. The national government had to
show its presence. The choice of locale for the first meeting was naturally the IFREMER center at Toulon–La Seyne. Presided over by the
Secretary of State for the Sea, Jean-Yves Le Drian, the installation of
the Coordinating Committee took on the air of a war council.

On January 24, 1992, the maritime prefect, admirals of naval headquarters, departmental prefect, local and regional directors of Maritime
Affairs, all in military regalia, had been assembled beside university
professors, ranking officials of IFREMER, local elected officials, and

94   persons such as Dr. Alain Bombard,[45] then a deputy of the European
Union. The press was also present. I was impressed by the sheer scale
of the event.[46] Everyone who had authority in the marine realm was
united against an alga that was decidedly a real nuisance. A sample of
Caulerpa taxifolia collected at Toulon was prominently exhibited. En-
throned in the middle of a large table covered with green velvet, the
accused was scrutinized by a crowd of curious onlookers, some stricken,
others smiling; it was surreal.

The Secretary of State for the Sea tried to lessen the drama of the
occasion; everyone could relax, the national government had decided
to take the matter into its own hands![47] "We must act quickly, but not
precipitously," he declared. He proceeded to the installation of the Coordinating Committee,[48] placed under the chairmanship of Hennequin,
interregional director of Maritime Affairs for the Mediterranean, and
he presented the two co-chairs of the Scientific and Technical Committee. He promised an immediate infusion of 500,000 francs so the proposed program could be initiated as soon as possible. He counted on
similar funding from the Ministry of the Environment. He suggested to
the local town councilors that the balance necessary for the research
program presented by the Scientific and Technical Committee (half the
sum) was the responsibility of local governments.

Boudouresque presented the research program. The representative
of the fishermen told the audience about the concerns of his organization and emphasized his satisfaction at seeing the Secretary of State for
the Sea react so quickly once he entered the scene. I spoke one more
time, on the urgency of considering the eradication of the alga. The
spokesperson for IFREMER, Merckelbagh, manifested some skepticism
about the scientists' arguments. He asked that an eradication be justified
by more clearly established risks and argued against any attempt using
chemicals. I had suggested to Hennequin that he invite Pierre Granaud,
the discoverer of the *Caulerpa taxifolia* colony situated not far from this
assembly in a Toulon inlet. He was invited and congratulated during
the session. Help from all persons who worked underwater would be

decisive in the struggle against the alga, and it was useful to emphasize
this collaboration yet another time.

Everyone seemed satisfied, though they expressed divergent hopes. Some thought they had succeeded in stopping the paranoia surrounding the killer alga by controlling the alarmist scientists, while others were persuaded that they had finally convinced the highest levels of the national government to take action. All hopes were to be rapidly dashed.

Diversions and Intimidations

February. The water was at its coldest, 12°C at the surface, 13°C underwater. It always takes a little courage to enter the gray sea, whipped by surf, with the rays of the sun too oblique to penetrate deeply. One first feels the slap of freezing water in the face and then the torture of its infiltration beneath the neoprene dive suit. Then one recovers while shaking a bit and is suddenly plunged into another world. I will never tire of the everyday features of life underwater: the behavior of a fleeing fish, the winter colors of multicolored algae, the mysterious clicks of molluscs, fishes, and crustaceans glimpsed through the turbulence of the two lines of bubbles emanating from the regulator.

The winter storms tear up or shred anything fragile between the surface and five meters; the *Caulerpa* "sink." The first fronds are shrivelled, broken, sparse. But at ten meters all seems calmer; a green carpet, thick and strong, covers rocks and sand. The alga is beautiful, it undulates under the attenuated influence of the swells. But I see too much of it, I have already seen too much of it. I see it every time I dive in this spot. I know that lower, to the right, to the left, it is the same. I know that I can swim over the *Caulerpa* prairies until I exhaust my air supply. It is everywhere. In three months, it will reawaken, proliferate, stretch out to cover and suffocate. I am enraged. Then I gather some, I shove it under the mask strap, under the weight belt, under the lifejacket, I sew a collar of it, I adorn myself with it, I am quickly encircled with garlands

96   of Polynesian ferns! I have them all over, I am completely green. My teammate imitates me. We resemble the French cartoon characters Dupond and Dupont who, in Tintin comic books,[49] have a green beard and green hair that never stop growing. Mockery in the face of our impotence against this supernatural prairie, so beautiful, so gentle, but also so deceitful, so cruel for the other algal prairies, for coastal underwater life of the Mediterranean. No, we are not stricken by dive narcosis, the terrible affliction that each year maddens and kills many divers; it is simply a collective letting off of steam on the subject that haunts our dreams. We bathe ourselves in *Caulerpa*, we roll around in it. We laugh, it is good for us. It is time to surface, to go through the stages of decompression, to return to reality. The green monsters rise from the bottom, pick off shreds of green flesh that sink in the current. Surface! It is now necessary to confront other hydras, other evils.

BRICE LALONDE, MINISTER OF THE ENVIRONMENT, COMES TO NICE

There has always been competition between the various ministries, especially on subjects where they all have jurisdiction and still more if the subject is hot and well publicized, thus capable of increasing their status. So I was not surprised to learn that the Minister of the Environment, Brice Lalonde, intended to show his concern for the problem posed by the invading alga one week after the engagement of his colleague, the Secretary of State for the Sea. Lalonde wanted to visit our laboratory to encourage us—and to take advantage of this trip by switching hats from member of the government, in an official visit to the Côte d'Azur, to leader of the political party Génération Écologie. He wanted to unify two local lists of green candidates who were squabbling over the coming regional elections. His cabinet hoped that the visit to the laboratories engaged in research on *Caulerpa taxifolia* would be followed by a presentation of the problem, in the presence of the directors of the agencies that conducted environmental research.

Jaubert at this time had one foot in the Faculty of Sciences at the University of Nice (where he still taught) and the other at Monaco (where he had just been named director of the European Oceanographic Observatory). The president of the University of Nice–Sophia Antipolis informed me that he intended to invite him. Six months previously, I had opposed Jaubert on the subject of *Caulerpa* before a reporter from *Science et Vie*, dumbfounded by the highly nonacademic spat. Jaubert held exactly the same opinions Doumenge had expressed to the media. I informed the president of the university of my profound lack of enthusiasm; by inviting him, he was taking the risk of provoking a fruitless argument that could disturb the decorum of the official visit.

Jaubert in the end was invited by the president and, just as predicted, took the floor to sow doubt about the toxicity of the alga and its environmental impact. In presenting himself, he did not acknowledge his functions at the Oceanographic Museum of Monaco, taking advantage of the fact that the European agreements had motivated the creation of the European Oceanographic Observatory that he directed, and stating that he was thus charged with studying major natural threats to the sea. He then exhibited what for him was proof that the alga is innocuous: a photo taken in a Museum aquarium of a Mediterranean fish with a fragment of *Caulerpa taxifolia* in its mouth.[50] He aimed thereby to illustrate one of Doumenge's preferred bits of sophistry: the alga is eaten by Mediterranean fishes, it does not contain toxins, and therefore it is not dangerous. This caricature of an argument, aimed more at the media than at science, proved nothing. Was the fish (a sea bream, an omnivore) nourishing itself with the alga or playing with it?

It was clear that Jaubert was attempting above all to defend the interests of his Monacan institution. But his opinion, expressed before an assembly of administrators, politicians, chemists, and physicists, took on the value of a carefully formed hypothesis. A title that sounded impressive and a simple photo solved the problem of the toxins that two teams of scientists, headed by Pietra and Pesando, were forced to study according to normal scientific rules. This ploy cast cold water on the audience assembled around the green covering of the immense table of the theater of Valrose castle, seat of the presidency. The Minister, feeling indirectly criticized for having come so far to deal with such a minor

98 problem, justified his interest in the phenomenon by arguing that it is easier to prevent such risks than to cure them. But Jaubert's opinion was seconded by a member of the Departmental Council of Alpes-Maritimes, invited by the president. Both strove to plead their arguments to the Minister.

I later learned that this effort had the expected effect; the Minister quickly cooled towards the problem after the dramatic spectacle of two university scientists arguing on opposite sides of the issue. This "salade niçoise" had to become clearer before the ministry could get involved. Throughout the ministerial tenure of Lalonde and that of his successor, Ségolène Royal, the funding promised to study or fight the alga was never advanced to the Scientific and Technical Committee.[51]

To justify his journey, the Minister had, of course, called on the press. Once again, the problem sprang to the public eye. The specific debate was not reported, as it took place behind closed doors, but the reportage on the arrival of the Minister mixed the struggle against the alga with political maneuvering. The goal of his visit was depicted in these terms: "Brice Lalonde between flora (marine) and landscape (electoral)."[52]

In February, I was sounded out about the possibility of joining several electoral tickets and participating in political meetings. I systematically rejected all these advances and never wanted to exploit my press notoriety. Obviously, the name *"Caulerpa"* could help those who were running in the regional elections. In the press, candidates took a position for the fight against the alga by proposing various steps. Many high elected officials felt concerned and drafted written questions for the Minister of the Environment or the Prime Minister.[53] Others lodged formal legal complaints.[54] Were these highly publicized steps sincere or opportunistic? In any event, all rapidly sank into oblivion or into the net cast by my opponents.[55]

Despite my strict political neutrality, a rumor began to circulate on the coast that I had joined certain local, radical Trotskyite political circles that had infiltrated the bourgeois society of the city. The killer alga reflected unfavorably on the Côte d'Azur, and I was fingered as a political agitator. From this period onward, the Departmental Council of Alpes-Maritimes halted the assistance it had granted to my laboratory. At the same time, the diving team, consisting of employees of the Departmental Environmental Bureau, were no longer supposed to concern themselves with the spread of *Caulerpa taxifolia.*

During this period, my energy was turned towards an internal struggle. My laboratory was immersed in a strange affair, as we had been steered towards opposing the laboratory directed by Jaubert.[56] This was during a time well known to researchers, when their work is evaluated. The results of this evaluation determine research funding, laboratory space, recruitment possibilities, promotions, indeed entire research directions. Everything was at stake. The verdict came in January 1992. My marine biology laboratory and Jaubert's had been evaluated by a committee of specialists in earth, atmospheric, and planetary sciences. The judgment can be summarized in a few sentences and surprised no one—our laboratory was a washout, his was very active. The two laboratories ought to join. This "advice" meant that I ought to fold my team into Jaubert's. Now, at the same time, another committee (of biologists) had evaluated my laboratory (and not Jaubert's); my laboratory was recognized as competent and qualified to be part of the honorary list of teams suited to supervise doctoral programs. Nevertheless, Jaubert suggested to me officially, in writing, that we join; he invited me to become part of his laboratory. This initiative had been supported *sotto voce* by the central administration of our university, which desired a rapprochement. Faced with our determination against this "hostile takeover bid," the presidential advisors conceded that our laboratory could remain independent and benefit from the university largesse that flows to a designated doctoral program for two years, at the end of which our work was to be reevaluated.[57]

This struggle for influence (typical in academia), for my honor, and for freedom of choice of my research subjects, at a time when all our energy was focused on the study of the invading alga, was followed by a war over laboratory space. At this time, our laboratory was very cramped. The administration had just redistributed the space of a for-

100   mer laboratory to the benefit of Jaubert's laboratory and of mine. A spa-
cious area, occupied until then by members of Jaubert's laboratory, had
been given to us. We desperately needed this space to pursue our research under proper conditions. Another office had been given to two
members of Jaubert's laboratory who, in any case, conducted all their
research at the Oceanographic Museum of Monaco.

It took us sixteen months and a lot of work simply to have this
administrative decision enforced. The space that we had been given
remained occupied by Jaubert's team, who became "squatters," so to
speak. This trench warfare between colleagues located on the same floor
and collaborating in the same courses was detestable. When we had finally been able to move into the space, this was grudgingly yielded to
us in a state of repulsive filth and graced with a big sign saying: "The
only true Taxifolia is Alex Meinesz, who expels colleagues to take over
their space!" A consummate, deplorable act of academic pettiness. The
story of *Caulerpa* began to resemble the televised soap opera *Dallas*.
Many other episodes were to follow.

Commander Cousteau Accused of having Introduced the Alga

A journalist from *Sciences et Avenir* conducted an inquest on the origin
of the alga. He wanted to know where the *Caulerpa taxifolia* cultivated
at the Museum of Monaco came from. After questioning the directors
of public aquaria, he was able to show that the exceptionally resistant
stock of *Caulerpa taxifolia* came from a public aquarium in Stuttgart,
Germany, the Wilhelmina Zoologischbotanischer Garten.[58] But the trail
of the alga stopped there. The heads of the aquarium were unable to
specify the date of its importation. It had been either brought to the
museum by an amateur aquarist or imported directly from tropical seas
through the commercial market in exotic species. It will never be known
if this stock changed during its culture in the aquarium, or if it had been
chosen in the sea by a perceptive aquarist. However, it is a fact that its
vigor and its striking adaptation to various artificial aquarium condi-

tions had been noted immediately by the Stuttgart aquaria. By means of exchanges of exhibition material, it was finally brought to France in 1982 or 1983, first to the public aquarium at Nancy, then to the Oceanographic Museum of Monaco, where it was cultivated to use as "litter" in raising clown fish.

Having learned all this in the newspapers, I was able to verify the facts easily by an exchange of letters with the directors of the cited aquaria. I also learned that I had not been the first to learn about the resistant stock of *Caulerpa*. One of the heads of the aquarium at the Oceanographic Institute in Paris, having identified the proliferating stock at some point in the middle of the 1980s, identified it as being *Caulerpa taxifolia*. In 1987, in a specialized journal, he praised its resistance to cold in this prophetic sentence: "Though of tropical origin, its acclimation to cooler seas is not impossible." [59]

In January 1992, the former assistant director of the aquarium at Monaco, Dominique Bézard, returned from Kenya after a disappointing experience in a center of export of tropical fishes to aquaria. It was he who had accompanied me in August 1989, during my first dive to the *Caulerpa* at the foot of the Oceanographic Museum of Monaco. To show him how the situation had evolved, I invited him to dive to the *Caulerpa* at Cap Martin. Shattered by the transformation of the bottom, he confided to me that its introduction at Monaco had not been at all accidental. According to him, detritus from the aquaria was often tossed out the windows overhanging the sea. *Caulerpa taxifolia* had simply been tossed into the sea. He also confided in Vincent Tardieu, a reporter, who published his remarks. [60] But that was not all. He had another, much more important thing to say; he promised to talk to me about it after settling certain problems with his former employers at the Oceanographic Museum of Monaco, including Cousteau and Doumenge.

Robert Kudelka, local correspondent of *Radio France Inter,* also wanted to learn what Bézard wanted to say. But Bézard could not make up his mind what to do about his secret, canceling his meetings at the last minute, saying one day that he was going to tell all, only to retract the promise the next day. During his long discussions with the journalist, he most often mentioned his personal disagreements with the last director of the Museum. Before speaking, he wanted to go to Paris to see Commander Cousteau.

A simple coincidence? While Bézard hesitated to unburden himself, Doumenge stated in a local newspaper, "It is difficult to establish a population of *Caulerpa* without turning to delicate planting operations of a certain scale. Two or three individual algae that get out of an aquarium would not suffice in the least to accuse us of having planted them intentionally!"[61] Implantation! I then remembered the first declaration of Doumenge after the discovery of the alga at Cap Martin, when he judged reports of its proliferation anecdotal but did not contest their origin. In a clarification addressed to a reporter from *Nice-Matin*, he had twice mentioned an implantation, describing that operation as very beneficial for the environment: "These algae have long carpeted many tanks of the Museum aquaria. . . . In this regard, it should be noted that, for several years, thanks to the implantation of *Caulerpa* at the foot of the maritime façade of the Museum, species . . . have returned in great numbers to inhabit this zone that they had previously abandoned. . . . Unfortunately, this implantation will doubtless be reduced to nil during the next winter that is a bit colder than the preceding ones."

The chief of the aquarium at Marineland at Antibes, Pierre Escoubet, representing the Union of Curators of Public Aquaria of France (UCA) on the *Caulerpa* committees, knew the divers of the Oceanographic Museum of Monaco well. When the means of introduction of the alga was raised at the second meeting of the *Caulerpa taxifolia* Scientific and Technical Committee,[62] he cut short the discussion by saying that, to him, the deliberate implantation of the alga "was the most logical" hypothesis.

I thought Bézard's revelations could make the situation healthier and dispel the ambiguity of the algal introduction into the Mediterranean. I begged him to tell the whole story. The journalist Kudelka agreed once again to meet him, but Bézard suddenly retreated into silence. At their last discussion, he informed the journalist that, because

of various types of pressure, he could say nothing. I was thus unable to 103
gather eyewitness testimony on the circumstances of this possible implantation. I regretted above all not knowing two specific points: the exact date of the introduction and the number of thalli (individual plants) tossed out the windows or implanted. These elements would have contributed to a better understanding of the demographic dynamic of the alga.

In previous publications and my doctoral dissertation, I had described how one can easily implant the harmless Mediterranean *Caulerpa prolifera* to revegetate muddy harbors from which that alga had vanished for over a decade because of a succession of cold winters. It sufficed to attach the stolons of a fragment of algae with hairpins planted in the mud or sand. Two or three hairpins along the creeping stem and the alga could no longer be dislodged by the surf. Michel Hignette, assistant director of the aquarium at Monaco, predecessor of Bézard and former president of UCA, had used my method in June 1983 to implant the innocuous *Caulerpa prolifera* in Monacan waters with the aid of another Museum diver.[63] The implantation site must be well chosen; it must not be exposed to swells and, to enable one to find the cuttings again and to observe their growth, it's better to plant them alongside a landmark. Now, the other *Caulerpa*, "*taxifolia*," was seen for the first time 12 meters deep, next to an intake pipe for seawater that fed the aquaria of the Museum.

Experiment gone awry? Small-scale attempt to culture an alga that was useful for the aquaria, conducted unbeknownst to the chiefs? Tossing out the window bits of algae growing too well in the aquaria? It matters little! I never told the news media about the spicy elements of this case. Although they would surely have had spectacular impact, I considered the question of willful or accidental implantation of *Caulerpa* to be of distinctly secondary importance. As a specialist in these algae, I would have gladly bet that the alga would not survive the first winter and would not grow. No one could have foreseen its spread or estimated the consequences of such an introduction.

The Museum scientists knew that the exotic alga cultivated in their aquaria did not exist in the Mediterranean. It was impossible for them to imagine what was going to happen. But instead of handling the problem when the situation began to become worrisome, those in charge chose to question its origin and to propound its merits. By denying the threat posed by the hostile proliferation of an introduced species, these authorities proved to be negligent and imprudent.[64] The vigor of the subsequent debates is doubtless related to this abdication of responsibility.

Upon Bézard's return from his "consultations" with Doumenge and Cousteau, I had a sense of foreboding about the implications of the latter personage entering the public debate. He reacted quickly by sending a member of his team to study the matter in situ. I received his representative, the oceanographer Denis Ody. With Bézard, we dove to the *Caulerpa* fields at Cap Martin. He immediately admitted the existence of a major threat. On February 2, 1992, before the Commander's name appeared in the press, I wrote to him. After recalling my participation in his book about the Mediterranean[65] and summarizing the facts of the *Caulerpa taxifolia* invasion beneath the Museum, I promised my support to him. In my capacity as specialist in *Caulerpa*, I informed him of the impossibility of having predicted the adaptation and spread of this tropical alga in the Mediterranean during the first years after its implantation, years during which he directed the Museum.

The algal implantation beneath the Museum was innocent at the outset and so anecdotal that he had probably never been informed of it. It was only after his departure, in December 1988, that the situation should have drawn the attention of the new directorate. It was in 1989 that the occupied area grew from several hundred square meters to nearly a hectare. I informed the Commander that I had not discovered the situation until August 1989, and that I had immediately informed the new directorate about it. Cousteau did not respond to my letter. But the arrival of his representative on the coast had been noticed by several reporters who had even attended the dive at Cap Martin. Inevitably,

Cousteau's name began to appear in the press. For more than fifteen days, he was directly associated with the affair. The way in which his name was used and the chronology of the citations are worth recalling.

On February 15, 1992, a national magazine asked: "On the Côte d'Azur, one question haunts the scientific community: is Commander Cousteau involved in what could become one of the greatest ecological catastrophes of the Mediterranean?

And if Cousteau is involved, what then? It has gone this far: this tropical alga, it is claimed, would have 'escaped' from the Museum of Monaco around 1984, when he was still director. The accusation is serious."[66]

Several days later, in response to a reporter from *Figaro* regarding the hypothesis of a Monacan introduction ("advanced by many scientists as well as by certain members of Commander Cousteau's entourage"), Doumenge stated: "Until December 31, 1988, Commander Cousteau was director of the Oceanographic Museum; it is thus up to him to answer this question precisely. I believe that the Cousteau Foundation, currently conducting research on this matter, will soon issue a clarification. Since January 1, 1989, under my direction, the Oceanographic Museum has progressively closed off the tropical tanks in closed circuits."[67]

Cousteau's representative, Denis Ody, lessened the Commander's responsibility in a very indirect way: "Who could have imagined for an instant that this type of alga could survive the winter, that is to say, resist temperatures between 11° and 13°C?"[68] But the strangest mention of the Commander was in a press release by the Museum directorate,[69] under the headline "A clarification by the Oceanographic Museum of Monaco": "As Professor Meinesz has stated that exotic algae have been introduced inadvertently in 1984 by the aquarium of Monaco, the administrative board of the Albert I Foundation wishes the following clarification to be known: Commander Cousteau, director of the Oceanographic Museum from 1957 to 1988, is too well known for his activities in defense of the marine environment for one to imagine that his attention would have been so faulty as to have caused the establishment that he directed to emit harmful or dangerous outflow. Moreover, it is easy to verify that, until 1986, no direct purging of aquarium waters to the sea took place, the elimination of overflow being achieved by pouring it into the collecting sewers of the city of Monaco."

This was too much! I decided not to let the attack pass and, above all, to appeal to reason: "It is necessary now, urgently, for all of us to work together to do what is possible to suppress this alga, if there is still time!"[70] Despite the report of the Museum press release in *Nice-Matin*, and my response to it, only one other newspaper reported on this episode, citing an argument between Cousteau and me: "On the origin of this destructive proliferation, there is a polemic between J.-Y. Cousteau and Professor Meinesz, of the laboratory of the marine environment of the University of Nice. According to the latter, it is likely that the *Caulerpa* came from the purged aquarium waters of the Oceanographic Museum of Monaco. The celebrated Commander, who was the director between 1957 and 1988, has formally denied this charge."[71]

On February 23, 1992, Cousteau was interviewed by a reporter from *Nice-Matin* who hoped he would quickly take a stand. He could get only an enigmatic commentary ("certain aspects of opinions stated here and there appeared to him to be false and extreme") and an announcement of a forthcoming clarification after assigning an investigator from his foundation to the case.[72] Despite the fact that I took a position firmly in his favor, I never received encouragement directly from him. This silence was noticed. In March 1992 the director of a divers' magazine was astounded by his silence: "It is shocking that Commander Cousteau, chief of the aquarium at that time, has not commented on this event that some do not hesitate to call an ecological catastrophe. It is true that it is easier to mount his favorite battle horse and accuse underwater hunters of all marine ills. However, do not the 'future generations' deserve this influential person's interest in this matter?"[73]

Having undertaken an investigation at the site and dived to the invaded sites, Ody published, in April 1992, a widely read article in the journal of the Cousteau Foundation, *Calypso Log*.[74] He adopted the views of the alarmists in emphasizing the threats the invading alga presented. Representing the editorial staff of the journal and Commander Cousteau, he gave his very diplomatic views on the subject of the origin of *Caulerpa taxifolia:* "*Caulerpa taxifolia* did not fall from the sky! The question of its origin has fed a prolific and vain polemic that we leave to the specialists. The first observation in the Mediterranean habitat was made at Monaco, and the scientists who know Mediterranean ecosystems are unanimous in their view that *Caulerpa* 'escaped' from an aquarium. Let us stick to these established facts."

In the autumn of 1992, a Cousteau biographer recalled all these facts and reproached him for his "only negligent act,"[75] that of not having dealt fully with the phenomenon when he learned about it. Cousteau always avoided entering the debate directly. Just once, in October 1992, did he minimize the problem, in a very unusual context. He was the guest of the Prince of Monaco, and, questioned during the Prince's reception, he declared that "it was an affair blown up out of all proportion."[76] It was not until January 1994 that he sent a letter (never distributed to the media) to the Minister of the Environment in which he described his great concern in the face of the algal invasion. Five months later, an article squarely in the alarmist camp appeared in the magazine of his foundation.[77]

While the press discussed Cousteau's responsibility, Doumenge increased his press statements, his only type of communication from January through March 1992. Soon he linked his Lamarckian hypothesis (a Mediterranean *Caulerpa* suddenly transformed by environmental change into another species) to the origin of the Mediterranean Sea. His new explanation was biblical, going back to the flood. He recalled that *Mare nostrum*, which dried up a very long time ago (5.7 million years), was first refilled with cold Atlantic waters; it then sucked in the warm waters of the Red Sea with its tropical algae, among which was *Caulerpa taxifolia*.[78] To buttress his reasoning, he even stated: "Two hours ago, a phone call informed me of its presence at Djerba."[79]

Sometimes he invoked the role of the greenhouse effect: "The general climate is warming and favoring the expansion of tropical species.

108   *Caulerpa taxifolia* is the bio-indicator of warming in the Mediterranean; if it has emerged from its dormancy in Monaco, it is because it has liked the warm and rich waters of the water purification plant of the principality."[80]

Sometimes pollution was the cause: "In my opinion, the growth of this alga is linked to zones rich in nitrates and phosphates; that is why it has a tendency to proliferate near sewer outfalls."[81]

Sometimes he challenged the identification of the alga: "No one has proven that it is a matter of *Caulerpa taxifolia* since no scientific description of this species has ever been done." "The alga has not yet really been characterized; is it truly *Caulerpa taxifolia?*"[82]

Sometimes he justified the presence of the alga in the Mediterranean as useful: "In fact, the Mediterranean is a tropical sea that lacks warmwater species; *Caulerpa taxifolia* produces enormous amounts of oxygen and reinforces that environmental equilibrium in the habitat it occupies. Moreover, we are going to prove this sci-en-tif-ic-ally!"[83] From 1990 to 1995, Doumenge published no scientific articles on the subject of *Caulerpa.*

Sometimes he reduced the debate to a personal quarrel: "The alarmist conclusions reached by some of his colleagues are not based on serious scientific study. He specifically denounced the attitudes of Professors Meinesz and Boudouresque: 'I deplore the fact that these young men adopt an irresponsible, unscientific attitude. When a solid database has been amassed on this subject, they will be ridiculed. The whole *Caulerpa* business is as clear as mud. This allows people to talk and 'especially to get research funding.'"[84] We argue "like ayatollahs," he complained in a national telecast.[85]

The media debate over the origin of the invasion had overtaken the arguments about the dangers of toxins. But the reporters understood the reason for this debate; they recognized that *Caulerpa taxifolia* came from the Museum and, in the following months, did not hesitate to discuss its appearance in the Mediterranean in a straightforward and unambiguous manner. Doumenge's stubborn defense of his hypotheses and his persistent statements about us manifested surprising self-assurance.

This led me to fear the worst: the "abandonment" of the alga by the authorities. Alas, my impression proved to be justified at the first meeting of the Coordinating Committee on *Caulerpa taxifolia.*

Nevertheless, It Does Grow!

They first made use of the ticking clock: the Coordinating Committee was convened a month after its solemn installation at Toulon. Such a slow pace in the face of a dynamic problem testified to an evident lack of enthusiasm for our ideas. Held on February 24, 1992 at Nice, in the premises of the Prefecture of Alpes-Maritimes, the meeting was a disgrace. Everything possible was done to humiliate the scientists who were charged with alerting the authorities. It was a truly medieval exercise.

Boudouresque and I learned at the meeting that Doumenge had been invited as an observer. Unavailable on that day, he sent a letter to the committee chair to lament his status as an observer, which would not permit him "to expose in a coherent and explicit fashion the very numerous complex problems raised by the object of these studies." I was unable to discuss matters calmly with my opponent without at least a show of conciliation and, at a minimum, a halt to his bellicose statements. No scientific debate was possible as long as we were being slandered and humiliated.

The interregional director of Maritime Affairs in the Mediterranean next distributed xeroxed extracts from publications by Pakistani scientists, which had been sent to him by the director of the Oceanographic Museum of Monaco. These old documents tended to prove that *Caulerpa taxifolia* contained no toxins and that it could be confused with another alga known in the Mediterranean, *Caulerpa mexicana.* In the eyes of French nonscientists, these could be very impressive documents (they were highly technical and in English). In reality, a close examination proved that they showed nothing. Why had not the director of Maritime Affairs submitted these documents beforehand to the Scientific and Technical Committee? If he did not have confidence in us, why had he not called for an international review, as we had so often suggested to him?

As for the IFREMER representatives, they distributed reports of their first two dives, carried out in February 1992 on sites invaded by *Caulerpa taxifolia*.[86] The impact of the alga on the habitat there was depicted as minor. The assessment of an IFREMER ichthyologist[87] who dove at Cap Martin summed it up thus: "The native communities (the *Posidonia* meadows) seem unaffected by the establishment of *Caulerpa*." He was not an expert in algae and had drafted his observations after only two dives (that is, less than one hour of exploration), to sites where, during the season he dove, *Caulerpa* began to be affected by the cold water of winter. These documents also should have been examined beforehand by the Scientific and Technical Committee, which would have been able to adduce contrary evidence and to criticize the simple and partial methods used to describe the situation.

In place of a rigorous scientific examination of the "new" elements introduced by Doumenge and the IFREMER experts, we moved directly to the testimony of the administrators and elected officials of the Coordinating Committee. Our fears were contested, doubt was cast on all our research. People sought to attribute economic motives to us. This scientific drumbeat in the media, was not it all just orchestrated by a thirst for research funds? Elected officials and representatives of Maritime Affairs questioned the choice of administrator of funds for research on *Caulerpa taxifolia*, who was none other than Boudouresque, president of the quasi-public association Groupement d'Intérêt Scientifique-posidonie (GIS-posidonie), charged with managing the research accounts. This choice had, however, been supported by the Ministry of the Environment, the Secretariat of State for the Sea, and by all concerned scientists, including those of IFREMER. A regional councillor asked if this association satisfied accountability requirements for the receipt of public funds, even though the association had existed for more than ten years and this same councillor had participated in votes awarding it contracts. The senior bureaucrats and politicians had, however, ample time to examine these technical questions before the meeting. The administrative documents of the association and the legality of the transactions could have been examined at any moment and were still available. Thus, beneath the surface of the proceedings, Doumenge's public statements about our motives dominated a fruitless media debate.

By contrast, no one cared about the delay in making the necessary funds available to the laboratories. Neither the Ministry of the Environment nor the Secretariat of State for the Sea had awarded the sums that had been promised in the press and were said to have been released. Not one cent was distributed by the national government. The only concrete action was the creation of a third committee on *Caulerpa taxifolia,* called the "Committee on Financing," charged with coordinating the requests for funding needed for research and to fight the invasion. In fact, the key goal of this committee was to control the allocation of funds that had been granted.

Finally, as the last aspect of putting the alarmists on trial, the committee produced a ruling on our freedom of expression. This nasty job could not be initiated by an administrator or a politician, so it fell to a representative of the fishermen, Aquila, whose turnabout surprised us. He called straight away for the censure of the scientists, wishing that only the chairman of the Coordinating Committee give statements to the press.[88] He called for the creation of yet a fourth entity to manage the invasion of *Caulerpa taxifolia:* a communications bureau with the exclusive duty of reporting on all aspects of the problem. Quite a change in tack for a man who had heralded a catastrophe just sixty days earlier. Had he forgotten our long telephone conversations, in the course of which I had explained the problem to him and had calmed his fervor? Had he forgotten his demand, broadcast to whoever would listen, for an administrative inquest to counter the activities of those who had been negligent in this affair?

It is true that his position as president of the regional committee of fisheries was at stake, as elections imposed by a new law had been announced. He was a candidate and must have felt rebuked by his electorate over the publicity about the toxicity of the alga; this was an electorate fearful above all that consumers of seafood would doubt the quality of the catch. To quiet the alarmists, who were seen as agitators hindering the sale of local fish, had become his new hobby horse. At the same time, he became a faithful collaborator with his administrative supervisors: Maritime Affairs and its institutional scientific advisors (the IFREMER experts).

The director of the cabinet of the Prefect of Alpes-Maritimes took over the task of insisting on the necessity of controlling communications. The following sentence was attributed to him in the minutes of the meeting: "The scientists cooperating in the activities of the committee should, if they are queried by elected officials or the press, first contact the chairman of the committee." At the press conference, Hennequin and Mrs. Andrée Heymonet gave only soothing statements.[89]

Although shattered by the thrust of the discussion, the "alarmist" scientists respected the wish of the fishermen, who were supported in their moves by the elected officials and the senior bureaucrats of Maritime Affairs. We therefore refused to explain ourselves in public. The effect was the opposite of what we had expected. Under the headline, "The mute worries of the scientists," the local newspaper reported that "nothing very definite was suggested for immediate action and there was a notable attempt to make the situation seem less alarming."[90]

The next day, I was astounded to learn that the interregional director of Maritime Affairs for the Mediterranean had arranged to distribute to reporters a status report composed by IFREMER.[91] Very one-sided, very reassuring, it included several major scientific errors. This "scientific status of the situation" had not been distributed to the members of the Coordinating Committee or examined by the chairmen of the Scientific and Technical Committee. This was all merely a pretext to control the flow of information; the eradication of the alga was going to be stopped and our studies could wait. The main objective of the report was to muzzle the alarmists in the media. To top it all off, I learned several days later that an anonymous letter had been sent before the meeting to several members of the committee and to the main political leaders of the entire region.[92] The letter treated us as frauds and criticized our wish to see the alga eradicated. During the meeting, no one reacted against this cowardly and vile action. Worse, I learned later that several bureaucratic departments studied the letter and sent it on for further consideration.[95] I made the connection between this letter and the fact that we were in the crosshairs at the meeting, subjected to deceitful questions. In this nauseating climate, research grants could only be curtailed.

Only two men stood up for us. Admiral Tripier (since deceased), coastal prefect and commander of the Mediterranean fleet, first informed me of the venom being spread about us. He encouraged me in a battle that he found despicable and felt would be difficult. In addition, Professor Nardo Vicente, of Saint-Jérôme University in Marseilles, who was a member of the city council of Marseilles, had been informed of the anonymous letter by the mayor, who had sought his advice. Denouncing this letter as a tawdry act, he wrote to the mayor that he had the utmost confidence in our activities and that he deplored the cowardly conduct of our opponents. From the other attendees and their subordinates, we received nothing, at best a mute reproach. Our detractors were apparently prepared to do anything to defame us. The lowest of blows were allowed.

I dreamed about giving up my role on the committees that had been created. The doctoring of information and harassment of opponents of the party line gave me the impression of being in a totalitarian system. My advocacy of the usefulness of international expertise was in vain. I was prepared to mount a counterattack in the media, but I was increasingly convinced that a last-ditch effort would not succeed in mobilizing the authorities to intervene before the approaching summer. Any response would be too late to control the alga.

An expert evaluation of the problem was still needed. My colleague Boudouresque dissuaded me from any desperate, extreme actions. He thought the situation could still be controlled, and he had complete confidence in the system established by the government. He promised to expose all the maneuvers that had targeted us, and, in fact, he did just that at the next meeting of the committee, in April 1992. The inter-

114   regional director of Maritime Affairs in the Mediterranean and the IFREMER technical experts were treated to a remarkable session of denunciation of the organized disinformation campaign. Boudouresque's presentation was so scathing that he was asked several times to show more restraint.

This was how the scientific legitimacy of the enterprise was to be laboriously reestablished. However, for want of political support, the experts made no progress. Spring of 1992 was here. My latest dives to the *Caulerpa* meadows had shown me that they had come to life again. Much tender green regrowth carpeted the low, scattered vegetation that had just overwintered. It was beginning all over again. The alga was moving on to conquer new territory.

The Stakeholders Squabble...
and the Alga Spreads

Beginning in February 1992, the associations of divers, fishermen, aquarists, and conservationists entered the debates. Elected officials and their advisors and experts staked out positions that led to total disarray, with no global strategy to deal with the invasion. Underwater, new actors entered the scene: eradication experts. The main concern of most of these experts was to appear in or to control the torrent of media reports. Throughout this period, the alga flourished. Other Mediterranean countries were alerted, and the discovery of *Caulerpa taxifolia* was to lead to as great an outcry elsewhere as in France.

These tumultuous years, however, were to see the slow victory of the rational over the irrational. Persistently hostile to our "alarmist" ideas, IFREMER eventually surrendered in the face of the evidence. In 1997, the invasion of *Caulerpa taxifolia* was a fact recognized by everyone. Paradoxically, people are becoming accustomed to this new perturbation of our environment. But the problem is by no means being solved—on the contrary. Every announcement that the alga has progressed recalls the deplorable earlier episodes of polemic, sensationalist news reports, and official disorganization. This moral wound will be endlessly reopened; it will never heal.

The Pressure Groups

THE DIVIDED FISHERMEN

The alga, described as a plague capable of overthrowing ecosystems and eliminating the food of certain commercial species, and burdened with a regrettable reputation of toxicity, should have mobilized the association of professional fishermen. In fact, their reactions have been impulsive and often contradictory.

To understand the fishermen's attitude in the face of the *Caulerpa* invasion of the Mediterranean, one must know the status of professional fishing in the department most affected by the invasion, Alpes-Maritimes. Fishermen there plied their trade over small coastal bottom patches using small, traditional boats. From the end of the last war, their number had fallen because of bad resource management. Increasingly modern on-board equipment (motors, winches, nylon nets) had led to the exhaustion of fish populations on the small accessible fishing grounds (the continental shelf is very narrow at the Côte d'Azur). The number of registers (lists of sailors comprising the crews of fishing expeditions, approved by Maritime Affairs) had fallen from 450 submitted in 1950 to 350 in 1970. From 1965 to 1988, rampant construction on the coast had significantly reduced their area of operation: 13 percent of the small bottom patches between sea level and twenty meters have been covered or diked on the coasts of Alpes-Maritimes and Monaco by the construction of sixty structures. Moreover, pleasure craft anchoring during the summer, in all the coastal inlets, hinder the setting of nets. Finally, the establishment of marked bathing zones on a large part of the coast prevents fishermen from entering a protected strip of the 300 meters nearest shore during daytime. The combined impact of these constraints is reflected by a further reduction in the number of fishermen: in 1990, Maritime Affairs for the Nice region (which includes the coast of Alpes-Maritimes) received only 125 registers.

But despite the reduced competition, the catch remained generally insufficient to support the fishermen. The profession is a disaster, the

fishermen comparing themselves to the "last Indians." The various restrictions placed on fishing have led many prud'hommes to demand
compensation for the absence of a large enough catch to support a liveli-
hood. Coastal cities and harbor-managing agencies subsidize some fishermen, directly or indirectly, by granting them extra docking spaces in
newly developed harbors, spaces they can rent out at a profit.

The sector located between Italy and Monaco, the prud'homie of
Menton, has no more than ten fishermen (bosses and sailors). Further
west, between Monaco and Nice, the prud'homie of Beaulieu–Saint-
Jean-Cap-Ferrat–Villefranche-sur-Mer comprises sixteen fishermen.
They rarely fish at Cap Martin and Cap d'Ail (the two areas most heavily invaded by *Caulerpa*) and prefer to catch migrating pelagic fish
(tuna, greater amberjack, sardines, anchovies) or to set their nets at
great depth (between 50 and 200 meters). From time to time, the fishermen encircle the capes with their nets to catch rockfish.

As soon as the Menton fishermen noticed the *Caulerpa* invasion,
two opposing factions formed. One clan (three fishermen in the same
family) felt that *Caulerpa* was not changing the catch in the vicinity of
the capes.[1] For other fishermen and their chief (the prud'homme),[2] the
catch of certain species was in decline.[3] Everyone agreed on one point:
Caulerpa taxifolia hampered fishing in the autumn and early winter.
This is the season when the *Caulerpa* density is maximal and storms
rip it up. As soon as heavy swells arise, nets are clogged with *Caulerpa*
fragments. It is impossible to unravel them, and the nets have to be
left in the sun for the enmeshed algae to rot, rendering this equipment
unusable for at least a month.

At Menton, the two men who came to office as prud'hommes after
the *Caulerpa* invasion had more than fifty years of fishing experience
in the region. They had observed the ravages of the alga and had often
reported them to the departmental office of Maritime Affairs, their
supervisory authority. At the end of 1995, they called for the study of
methods to eliminate *Caulerpa* trapped in their nets. As for the fishermen reporting to the prud'homme of Beaulieu–Saint-Jean-Cap-Ferrat–
Villefranche-sur-Mer, who were much less affected by the invasion until

118   1996, they had been vigilant. As soon as several of them found algal fragments in their nets, they alerted the departmental office of Maritime Affairs.[4] As we have seen, their spokesperson and prud'homme has strongly supported the "alarmist" ideas. Divided, few in number, and more preoccupied by the day-to-day results of their fishing, the other fishermen of Alpes-Maritimes and of the Department of Var—in which the fishing sectors were distant from or still only slightly affected by the first worrisome signs of the invasion—were unconcerned by the alga. The soothing statements of the scientists from the Museum and IFREMER (from 1990 to 1992) had even reassured some of them.

I have described how Augustin Aquila, president of the regional committee of fishermen, got busy after our meeting in Nice in December 1991. The meeting he organized shortly thereafter at Toulon attracted many fishermen from Var, Bouches-du-Rhône, and even Languedoc-Rousillon, as well as bureaucrats who came to show their interest in the alga. The fishermen demanded immediate action and an inquest on the origin of the alga and on the shortcomings of the authorities who had been in charge of the situation. A month later, there was a sea change in strategy. The debate in the media on the toxicity of the alga had rendered seafood suspect in the eyes of consumers.

Pragmatic, and observing the harm rumors about the alga were causing to his industry, the regional leader of the fishermen found scapegoats among the scientific alarmists, whom he tried to silence. He rallied around the hypotheses of Doumenge and of his supervisory ministry, which were those of IFREMER. The administrative inquest demanded in December 1991 was consigned to oblivion. His moves were facilitated by the visit of the Secretary of State for the Sea, who made him a key figure and arranged his participation in various commissions dealing with the subject. Several months later, at the election for the regional committee of fishermen, Aquila was handily reelected, even though a battle was warranted.

The president of the committee on fisheries of the Department of Alpes-Maritimes, Alex Plusquellec, strove equally to participate in some meetings of the Coordinating Committee on *Caulerpa taxifolia*. Quickly perceiving the absence of any real will to come to grips with the alga, he made a striking declaration that his participation in the "masquerade" was futile. The situation led to rapid opposition between the fishermen of Alpes-Maritimes and those of the rest of the region. Only the most affected fishermen (those of Menton) used the media routinely to publicize the damage caused by *Caulerpa.* The president of the regional committee, Aquila, went so far as to belittle the worries of certain fishermen of Alpes-Maritimes, saying that the fearful fishermen "are, in reality, sport fishermen seeking subsidies."[5]

Far away, the fishermen of Corsica, whose sensitivity in all matters concerning their patrimony is well known, made several menacing statements, publicized their worries,[6] and announced the filing of a formal complaint against the Oceanographic Museum of Monaco for the risks posed by the eventual arrival of the alga on the "Beautiful Island."[7] A reporter received even more threatening suggestions[8]—the alga had better not reach Corsica. By 1993, the menace was nearing. A colony of *Caulerpa taxifolia* had grown in front of a sheltered bay on the island of Elba, less than 65 kilometers from Corsica. But by December 1998, the alga had yet to be found in Corsican waters. Many diving clubs, students from the Fesch school of Ajaccio, researchers from the University of Corte, representatives of the environmental agency on the island, and divers of my laboratory have been active in mobilizing the sea-loving public to keep close watch on anchorage areas of the "Beautiful Island."

THE MOBILIZED DIVERS

New colonies of *Caulerpa taxifolia* have always been found in locations between three and ten meters deep: not deep enough for the colonies to be mapped by boats equipped with retractable video cameras, but too deep to distinguish from an airplane or helicopter. Direct visual observation, by divers, is the only alternative left.

Our team, composed of a dozen experienced divers (colleagues, students, and external collaborators), could not aspire to monitor the alga

120   on the entire Mediterranean coast of France. We needed to solicit the assistance of thousands of divers and underwater hunters who explore the seafloor year round. These divers are my allies; I have been a member of the powerful French Federation of Underwater Sports and Research (FFESSM) since I first started diving in 1967. For several years I was even vice-president of the Côte d'Azur committee encompassing a hundred clubs. FFESSM claims more than 150,000 licensed divers as members, and affiliate clubs are present in all coastal cities. The exploration of the seafloor is not only a sport; it has become a tourist activity. Every summer, a flotilla of several hundred boats belonging to clubs and private individuals cross French coastal waters, carrying groups of divers from all over Europe. In the Department of Alpes-Maritimes, the total economic activity associated with underwater exploration has for several years exceeded that of professional fishing.

    Beginning in April 1991, I broadcast appeals to divers to report all discoveries of the alga. All clubs on the French Mediterranean coast received our brochures and our posters describing the alga. Nearly 40,000 copies of three successive editions (1991, 1992, 1993) were shipped or distributed directly at diving sites by members of my laboratory. I had also written review articles for the FFESSM magazine *(Subaqua)* and other diving magazines. By bringing the scientists information essential for mapping the spread of the alga, club divers and those of FFESSM played a key role in the *Caulerpa taxifolia* affair. At the end of each year, a summary of the invasion status was published. These summaries cited all of our informants. Thus, little by little, we wove a veritable network of sentinels.

    The director of FFESSM, André Védrines, wanted to engage his federation more directly in the campaign. At the fishermen's meeting on December 21, 1991, he offered the authorities his help and that of 150,000 licensed divers—300,000 eyes to scrutinize our coastal seafloor. On February 21, 1992, at the first meeting of the Coordinating Committee, Védrines offered to organize a large research effort on the alga, in which each club would be assigned to monitor a portion of the coast.

He asked the interregional director of Maritime Affairs for the Mediterranean for a modest stipend to cover the additional costs that divers
would incur (meals for divers or gas for boats). He was granted this stipend on the spot by the interregional director, Hennequin, who also
served as chair of the Coordinating Committee on *Caulerpa taxifolia*
and chair of the Finance Committee.

I advocated an autumn diving campaign, when the water is still
warm, when the alga is the most visible, the densest, and when the
newly dispersed colonies begin to enlarge. But the divers were eager to
act immediately, while the "killer alga" was on the front pages. We always envisaged a vast campaign of total eradication. For that, we needed
a more precise idea of the spatial extent of the alga before it began
growing again. This last argument was decisive in choosing the observation period. There was wide interest in my organizing this campaign,
which should have covered all 500 kilometers of the coasts of the Departments of Var and Alpes-Maritimes. With few new discoveries to report, I had underlined the importance of negative observations—that
is, records of the absence of *Caulerpa* at a particular date in a narrowly
defined site.

We decided to have all the clubs dive on Sunday, April 26, 1992.
That day, the weather was not good: rough, cold waters, high seas. More
than fifty clubs, mobilizing more than 500 divers, inspected small
patches of bottom from five to fifteen meters deep, each in a specific
zone. I joined twenty divers from my club in the hunt. Everyone had to
traverse one kilometer of coast, accompanied by a team-member swimming five meters away. A distributor of diving gear gave each club, free
of charge, several diving vests featuring a drawing of a *Caulerpa taxifolia* frond inside a red circle with a diagonal bar across it: Stop
Caulerpa!

The strong participation of the clubs and the enthusiasm of their
members made the operation a success. It was well publicized by the
directors of FFESSM, who demonstrated their ability to mobilize a campaign. Charles Josselin, then Secretary of State for the Sea, sent a

congratulatory letter to the directors, which appeared in the federation magazine.[9] There was again talk of the "killer alga" monitored by divers.

Subsequent activity was more disappointing. The observations gathered during the dive were supposed to be transmitted to the head of the biology committee for FFESSM in each region, who was then supposed to communicate them to me. The same individual was also in charge of distributing the funds from the stipend. But the stipend was never paid, and the federation had to cover part of the costs. The heads of the clubs and the participants who had given their time, and who then had to pay out of their own pockets, were disappointed. I received no reports on the inspected zones. The heads simply reported to me that few clubs had found the alga; the colonies that were observed had already been catalogued by our laboratory.

The French Navy joined the operation. Combat swimmers and divers among the crews inspected the coast in zones where military bases were established and where diving was prohibited: the island of Levant, Toulon harbor, and several other smaller zones. I received a precise report of the zones that had been surveyed, and no *Caulerpa* was observed. The only blot on this otherwise exemplary action, which could have been chalked up to training of military divers, was the exorbitant cost of the activity, tallied as 1.753 million francs! This activity allowed Hennequin to include this figure in the expenses borne by the national government in the battle against the alga—a clever way to hide the total absence of national funds released for the planned research.

Despite all these annoyances, the absences recorded during the coastal inspection allowed us to confirm the map of the algal spread established by our laboratory, thanks to the sustained vigilance of the divers who had been sensitized by our brochures. Our maps were reliable; we could base our eradication strategy on a precise assessment of the area occupied by *Caulerpa taxifolia*.

Not all divers experience the problem of *Caulerpa* in the same way; their degree of awareness depends upon their proximity to the invaded zones. The clubs at Menton, Roquebrune–Cap Martin, Beausoleil–Cap d'Ail, Beaulieu, Nice, Villefranche-sur-Mer, Cagnes-sur-Mer, Saint-Raphaël–Agay, and Hyères have been very active. Their directors dove at Cap Martin to familiarize themselves with the situation. They became excellent conduits of information to their divers. There has been less activity by clubs with territories far from the affected zones. Some have even remained skeptical because of the constant media drone about an alga they have not yet seen in their sectors.

Cap Martin is the favored site of four dive clubs based at Roquebrune–Cap Martin and Menton. At the tip of this cape, along a kilometer of coast, are marvelous underwater cliffs with a very diverse fauna and flora. Two sites in this location have even been listed in successive editions of the catalog of the 200 most beautiful diving sites on the French Mediterranean coast.[10] At the tip of Cap Martin, a statue of the Virgin has been erected underwater, attached to the foot of the cliff, its face directed towards the multicolored wall. Hundreds of divers come each summer to visit the Virgin and the wall. In three years, *Caulerpa* has progressively invaded everything: no more sea fans, no more coralline algal concretions, no more flamboyant bryozoans; henceforth, a fluorescent green drape covers everything. The chiefs of the dive clubs of this zone were the first to notice that their activities were becoming more difficult. The trip organizers have to go further and further to explore sites that represent the floral and faunal richness of the Mediterranean bottom. The invasion has thus begun to affect diving tourism.

THE AQUARISTS BECOME INVOLVED

There are nearly a million aquarists in France. These are thoughtful persons who greatly enjoy watching silent animals in their aquaria. The beauty of fishes of rivers and of tropical seas, the technical mastery of a system of aquaculture that is sometimes very complex—these things fascinate and excite them. Tropical marine aquaria are much less widespread than freshwater aquaria, because to procure and to keep alive the jewels of the tropics (clownfishes, butterflyfishes, multicolored surgeonfishes, etc.) is more complicated, more technically demanding, and more expensive. Algae and invertebrates are also imported to constitute tropical decor.

This "aquarium network" is clearly the most plausible explanation of the presence of the resistant stock of the tropical alga *Caulerpa taxifolia* in the Mediterranean. We have noted the reporters' investigations that brought to light the use of this stock of exceptional beauty, resistance, and vigor at the beginning of the 1980s. It was first disseminated in the large public aquaria. In 1982, it reached the tropical aquarium at Nancy, where its ability to purify the water and even its nutritive qualities for some tropical fishes were appreciated. Jean Artaut, then chief of the aquarium at the Oceanographic Institute of Paris, also acquired some and praised it highly.[11] In the meantime, it arrived at Monaco (by 1982 or 1983), and the heads of the aquarium of the Oceanographic Museum began to exhibit it in the display tanks; they also grew it in their stock tanks. During a meeting of public aquarium curators, one Monaco director emphasized its qualities and was happy to have a reserve of this alga in the sea beneath the Museum. By 1984, the Museum divers knew that the stock of *Caulerpa taxifolia* was "implanted" beside the seawater intake pipe. They knew that the alga survived the winter and that it was spreading.

In 1990, as soon as people knew that they could gather armfuls cheaply in the Mediterranean, retailers began to receive consignments collected at Cap Martin; commerce in the species grew, with a price of thirty francs per "sprig."

In December 1991, after our informational meeting, the aquarist network would be directly questioned in this affair. Aside from the directors of the Oceanographic Museum aquaria, other curators of public aquaria entered the public debate. Two of them, Pierre Escoubet of Marineland at Antibes and Jean-Michel Chacornac, scientific director of the Marseilles aquarium (Marseilles Aquaforum), were especially active and were even absorbed into the *Caulerpa* committees established by the ministries. Both of them accepted the Monacan origin of the alga in the Mediterranean and the role of the aquarium network.

Escoubet, founding member of the Union of Aquarium Curators

(UCA)[12] and representative of that association on the *Caulerpa* commis- **125** sions, knew the managers and divers of the Monaco aquarium well. At a meeting of the Scientific and Technical Committee on *Caulerpa taxifolia* (April 13, 1992, at the University of Nice), he downplayed the collective responsibility of professional aquarists, and especially that of the Monacan directors, in stating that the most probable origin was not careless disposal into the sea but "deliberate implantation" (it seemed to him most logical that the alga had simply been planted experimentally). He had, moreover, often repeated to the press (just as I had) that eradication was urgent,[13] and he took an active role in studying eradication techniques.

As for Chacornac, he had quickly sensed the danger posed by movement of more and more kinds of organisms in the aquarium trade, stressing the laxity of relevant legislation and the difficulty of controlling the movement. In early 1992, he sent a note to the directors of all public aquaria and to members of UCA; it was also reprinted in an aquarists' magazine under the title, "*Caulerpa taxifolia:* the alga that reminds us of our responsibility."[14] To set an example, he proposed to eliminate all *Caulerpa taxifolia* from public aquaria.

Other aquarium directors took positions in defense of their profession. Far from the invasion battlefront, they defended the Museum (and public aquaria erected on the coast) or complained (with some justification) of the excesses of the press. The director of the aquarium of Trouville-sur-mer (Normandy) opined in an article in a dive magazine,[15] in which he mixed his feelings (justified) about the press exaggerations with his personal hypotheses on the origin of the alga (diminishing the responsibility of the Museum) or on the causes of its proliferation (pollution). The director of the Brest aquarium (Brittany) followed the same reasoning: strongly denouncing the press excesses, then recalling the affair of the suspect coral substrates at Saint-Cyr-les-Lecques, which allowed him to affirm that the Museum was not the only party responsible for the dissemination of the alga. He then took up as his own hypotheses those favored by Professor Doumenge, concluding by way of a joke: "*Caulerpa taxifolia:* a stroke of good luck for the Mediterranean!"[16]

It was above all after the discovery of the chunks of dead coral covered by *Caulerpa taxifolia* in the harbor of Saint-Cyr-les-Lecques, in January 1992, that the aquarists were fingered. A regulatory commission was created, and it promulgated a decree forbidding the sale, transport, or possession of the alga.[17] Officially accused and punished, the aquarists handled the matter poorly.[18] A petition against the decree appeared in an aquarists' journal.[19] The petitioners attributed so much importance to the use of this alga that they went so far as to claim "that this decree is no more nor less than a threat to marine aquaria."

In certain public aquaria, the alga was ripped out; in others it remained. Elsewhere, the label *Caulerpa taxifolia* was replaced by the names of related species: *Caulerpa mexicana* or *Caulerpa sertularioides*. Few policemen, responsible for enforcing the law, had been trained to recognize these "marine lettuces." It is true that the decree was useless in cities far from the sea and also that the accidental dumping of marine aquarium contents was most likely not the cause of the algal dissemination in the Mediterranean. However, the decree had a symbolic value; legislation had most importantly pointed to the most probable misdeed and had enacted a sanction against it. By the same token, it guarded against the risks of uncontrolled jettisoning of aquarium waste.

Caulerpa taxifolia was to be discovered in aquaria on the other side of the world. In August 1993, one of my colleagues of the Nice branch of INSERM was scheduled to speak on the toxins of this alga at a congress of specialists in Hawaii. I asked him to bring back a living stock of *Caulerpa taxifolia* from this region of the northern Pacific so I could cultivate it in my laboratory and compare it to the Mediterranean stock. Despite his greatest efforts and the assistance of Hawaiian phycologists, he did not find the alga in nature. By contrast, he found a specimen with exceptionally long fronds in a local aquarium. In Nice, it was observed to have the same characteristics as the alga that grew in the Mediterranean: it was resistant to cold!

In the autumn of 1994, a similar discovery worried us more. At the University of Nice we received a delegation of university oceanogra-

phers from the University of Tokyo, to whom we presented the story of
the invasion by *Caulerpa taxifolia.* One of the professors present recognized the shape of the alga; he remembered having seen it in a public aquarium in his country, where a tropical aquarium for corals had
been installed that had been perfected by Jaubert and inaugurated by
Doumenge. The director of the aquarium confirmed to him that she had
indeed been cultivating *Caulerpa taxifolia,* and he warned her of the
risks of disseminating the alga in the Pacific a few dozen meters from
the aquarium. In November 1995, invited with a delegation of French
colleagues (including Boudouresque) to lecture at the University of
Tokyo, I had a chance to visit that aquarium.

Enochima is a beach resort town of the Kanagawa Prefecture, in
a large suburban area of Tokyo, on the Pacific but nevertheless 6,500
kilometers from Hawaii. At the entrance to the aquarium, we experienced strong emotions—a large hexagonal aquarium filled with magnificent corals was framed by two tanks three quarters full of *Caulerpa
taxifolia* sheltering several multicolored fishes. Above the aquarium
with the corals, one could read on a bronze plaque: "Aquarium Monte-
Carlo, Principality of Monaco." Under the plaque, a letter with the
heading of the Oceanographic Museum of Monaco had been reproduced
with the following text in French (and its translation in Japanese): "We
are deeply grateful to Madame Yukita Hori, director of the aquarium of
Enochima, and all the persons thanks to whom this balanced aquarium
was able to be completed. We hope that, thanks to this example, this
procedure can be diffused throughout the world; we hope thus to be-
queath to future generations magnificent coral reefs whose environment we have preserved. September 1993." Signed: Professor François
Doumenge, director of the Oceanographic Museum of Monaco.

On one side of the tropical tank were printed all the explanations
of the procedure that had been used (Jaubert's patented Micro-ocean).
An aquarium employee explained to us that the project had cost nearly
100 million yen (about 2.5 million francs). He explained to us that the
Caulerpa had been grown at Enochima for one year before the opening

128   of the "Monte-Carlo" aquarium. He even remembered Jaubert's surprise at discovering the alga at Enochima when he came to supervise
the construction of the tank.

Where did the alga come from? The inquest had barely begun. The
researchers of the Oceanographic Institute of Tokyo had already scoured
their national professional aquarists' journals. In the most recent, one
could often make out *Caulerpa taxifolia* in photographs of tropical exhibition tanks. In winter, the water of the Pacific in the southern part of
Japan, starting at Tokyo, does not go below 10°C. In this region there is
therefore a risk that our Japanese colleagues are going to try hard to
measure. If the alga is found in aquaria on Pacific coasts, there is every
reason to think that it ought to be found in aquaria located on other
shores, in other countries.

THE SOCIETIES FOR ENVIRONMENTAL PROTECTION ARE MARGINALIZED

The algal invasion was a delicate matter to deal with in the framework
of community associations. It was a question, on the one hand, of a scientific problem made complex by the highly publicized polemic and, on
the other hand, of an assault on a habitat that the majority of members
of these associations never saw. The radicals and eco-politicians had little interest in getting involved with a problem in which there were not
well-defined adversaries—no industry or pressure group was implicated, no obvious political misdeeds. The subject was hardly worth it.

I hesitated to intervene to mobilize this sector. It would have been
very easy to do, because, since 1973, I had lobbied associations for nature
protection in the Provence–Alpes–Côte d'Azur region. My volunteer activity in this movement was limited essentially to scientific advice on all
matters concerning threats to the marine environment, which had been
so mistreated during the last few decades by coastal construction. As
chair of the scientific committee of the Regional Union for the Protection of Life and Nature (URVN), an umbrella organization that encompassed nearly two hundred associations in the Provence–Alpes–Côte
d'Azur region, I had since 1984 been appointed to the regional economic

and social Council, succeeding Alain Bombard in this post.[20] I had in-   **129**
sisted to the successive directors of this powerful umbrella organization
that they should not make me wear two hats in this affair. I did not want
my scientific credibility tainted by my being labelled an eco-tactician.
Routinely informed of my battle, URVN respected my wishes and always supported me. The associations of the region most affected by the
invasion had often spoken out about the risks posed by the alga, but with
moderation and without making me a prominent part of their statements.

Nevertheless, the president of URVN, in the autumn of 1996, addressed an open letter to the Prime Minister, the Minister of the Environment, Agriculture, and the Sea, and the Minister of National Education and Research to ask them to take a stand on a new, fantastic
recrudescence of this affair—a dubious attempt by Monacan researchers to prove that the invasion of the alga was natural (an attempt that
became the object, as we will see, of an article in one of the journals of
the Academy of Sciences and of a rebuttal in the same journal). Three
months later, a single response was forthcoming: the Minister of the
Environment, Corinne Lepage, said that she lacked jurisdiction over the
ethical aspects of the problem but was very concerned with the spread
of the alga, and she announced that an international scientific seminar
on the alga would be held at the French Academy of Sciences in mid-
March 1997.

Other societies supported me by independently taking the initiative
of defending my hypotheses. The Friends of the Natural Park of Corsica
publicized its concerns, without compromising the two chairs of the scientific committees of the two natural marine reserves of Corsica, Boudouresque and me.

At the committees and commissions on *Caulerpa* established by the
national government, I was surprised to run into a Greenpeace official,
present as representative of nature protection associations. This international association had succeeded in establishing its status as observer of
official discussions. But the directors of Greenpeace must have quickly
noticed that it would be difficult for them to make their presence felt,

because the fierce combat among the various factions of scientists, government experts, and professional organizations occupied the field and the headlines. The Parisian representative of Greenpeace France, impressed by the threat the alga posed to the Mediterranean, presented a precise analysis of the situation and a perceptive critique of the government failure to come to grips with the phenomenon, in terms that I would have been glad to use.[21] During the autumn of 1992, with the Greenpeace ship *MV Sirius*, he participated in a mission to research the alga at Ibiza, in the Balearic Islands.[22]

Other groups became interested in the debate, usually seeking to inform their members without taking a position. The algal problem thus became the object of long articles in nature magazines,[23] several of which I was asked to write.[24] In an ecological newsletter, an analysis, very hard on the government, exposed in minute detail the incompetence of the people in power, denouncing their failure to attend to the file in a reasonable length of time and to try to control the alga.[25]

Among the local societies that reacted to the situation, I should mention two big civic service organizations, the Rotary Club and the Lions Club. Generally involving themselves in charitable works, they also participate in numerous activities benefiting the environment. The Rotary Club often invited Doumenge to give his views.[26] Some local Rotary Clubs, in order to be objective, invited me after he had spoken and published an article in their international journal.[27]

In the Lions Club, they were not content simply to listen and to try to understand; as a member I was able to organize my local Lions Club (Nice-Doyen) to sensitize yachtsmen about protecting marine life. To that end, we planned and produced a well-documented brochure on the riches of Mediterranean marine life and on all the concrete actions undertaken to defend it. Produced as a volunteer activity, printed thanks to the support of several publishers, ten thousand copies have been distributed free every year since 1992 by club members in all the harbormasters' offices of ports and harbors on the Côte d'Azur and in Corsica. Among the fifty pages of this pamphlet, the three devoted to *Caulerpa* send a very strong message: "If you haul up your anchor with pieces of *Caulerpa*, remove them right there or collect the *Caulerpa* *taxifolia* to dispose of them on land in your bag of household garbage, which will be incinerated. Don't spread *Caulerpa* from anchorage to anchorage!" Although it is impossible to evaluate the impact of this positive activity, this message, widely publicized since 1990 to those who are capable of carrying the alga from bay to bay, probably impeded its dissemination.

The "Eradicators"

Does it have to be eradicated? For the first six months of 1992, this was the first question asked at every meeting. The authorities thus gave the illusion of getting the problem under control. In practice, they were using the opportunity to delay making their decision.[28] At the end of the summer of 1992, all the players agreed on one thing—it was too late!

Time is the overarching factor in any struggle against a dynamic threat: sooner or later, the chances of success plummet and the costs skyrocket. It is a question of common sense. I did not spare my efforts to emphasize this point. To make everyone aware of responsibilities, I described in writing the importance of a rapid response and I proposed a strategy to try to finish off the alga before a new season of proliferation began. But neither the Prefect of Alpes-Maritimes[29] nor the authorities on the *Caulerpa* committees and commissions[30] discussed the proposal.

It must be said that the few preliminary eradication attempts were disappointing. In October 1991, a student and I had succeeded in clearing two square meters by hand in one hour. In February 1992, an IFREMER expert, assisted by five military divers of the Underwater Intervention Group (GISMER), achieved a similar result at Toulon, manually eradicating one to two square meters per diver per hour.[31] These first attempts showed that ripping the alga out by hand had to be done carefully. It was important not to let fragments escape because, once dispersed by currents, they could attach further away and initiate new colonies. This risk of inadvertent spread was communicated to all club divers who volunteered assistance. Our alert was relayed by the professional fishermen of Agay: "For goodness sake, do not do anything rash. No one should move on his own initiative until the scientific community has given precise directives that it will be able to apply and to control."[32]

It is true that the paranoia engendered by the "killer alga" and the prospect of a mobilization of divers for a complete eradication had generated many good intentions. Many divers volunteered to participate in ripping out the alga. The directors of their federation had tempered their eagerness by recalling the law: it is forbidden to harvest anything at all during a dive. The participation of the dive clubs could therefore only be realized if the national government asked for and directed it.

The dive club of Roquebrune–Cap Martin secretly tested a method that seemed logical—that of covering the *Caulerpa* with a sheet. By preventing photosynthesis, the divers hoped to destroy *Caulerpa* quickly. In January 1992, they installed six square meters of a type of black plastic commonly used in agriculture over a *Caulerpa taxifolia* prairie nine meters deep. This sheet was attached to the bottom with stakes, cinderblocks, and chains.[33] A month later, the divers found that winter storms had torn away the sheet and the *Caulerpa* were thriving.

At the first meeting of the Scientific and Technical Committee (in January 1992), the eradication project file had been assigned to the IFREMER experts. The mission of the latter was to review the various methods and to evaluate their efficacy and cost. This assignment was logical, as everyone had a role consistent with his or her status: university scientists did research, while the IFREMER experts were responsible for testing the feasibility of the eradication methods.

But in February 1992, before the tests had begun, the chief of the eradication project for IFREMER stated to the press that he did not believe in it.[34] For him, the attempt was not justified. Convinced that the alga would disappear of its own accord, Doumenge was of the same opinion.[35] In the spring of 1992, I had also ceased to believe in it, but for other reasons. The skepticism and the procrastination, expressed at the first Coordinating Committee meeting (in late February 1992), was unfortunately reinforced by the absence of financial support from **133** the national government. While the representatives of the ministries pressed the scientists to provide answers (Is it really toxic? Will it spread once again? Is it really eliminating the essence of the native flora?), at the same time they prolonged the process of getting research funding. Worse, they went back on promises made in the press, informing us of a decrease in total research aid. The message was clear: there was no political will to take charge in fighting the algal invasion. The summer arrived, the *Caulerpa* grew; there was no longer enough time to remove them all.

For want of an eradication campaign against *Caulerpa taxifolia* underwater, and given the funding squeeze organized within the official commissions, exacerbated by an absence of administrative decisiveness, the alga should have disappeared from the news. But the news media were watching. Two months after the creation of the *Caulerpa* commissions and committees, and after the reassuring visits of the Minister of the Environment and the Secretary of State for the Sea, the affair continued to intrigue the press. It was in this context that an unexpected attempt at eradication became a big story. It was conducted on March 3, 1992, by the town council of Saint-Raphaël, a community in eastern Var, at the base of the Estérel massif. I was far from the scene of this new development.[56]

THE MAYOR OF SAINT-RAPHAËL, ERADICATION PIONEER

The alga had been found in the autumn of 1991 in a small, private, sheltered harbor at Agay, part of the town of Saint-Raphaël. There were then no more than five square meters of *Caulerpa taxifolia*, which mayor René Georges Laurin wanted to destroy before it spread. He had begun by sending Minister of the Environment Brice Lalonde a formal request: "Since the only answer to this algal invasion is manual eradication, we ask for government support. . . . It would please me greatly if the action plan that you surely have not failed to draw up to battle this alga could be applied as quickly as possible. If this assistance cannot be

134 provided within a month, I will have to have the town take charge of the operation."[37] In the absence of a response from the minister, the mayor had had a third of the *Caulerpa* removed by hand by his municipal firemen. Another third was covered by a large plastic sheet attached by metal wire netting. The last third was spared to see what would happen to it.

Reported to the media, this second experiment with an opaque sheet aroused high hopes even before the method was proven effective. The operation made the front pages of local newspapers.[38] The mayor was able to hype the value of his action and force the Minister of the Environment to pay attention to it: "If the national government fails to deal with this matter, coastal communities will have to take charge of this type of operation, in order to prevent the reproductive period that begins in the spring from dramatically aggravating a situation that is already worrisome."[39]

It was just twenty days before the cantonal and regional elections. The mayor, who was also a senator, his environmental assistant, and a member of the departmental council ran as pioneers in the battle—finally, a group to set an example! Everything had been organized, from creation of a "municipal *Caulerpa taxifolia* committee" to publication of a sheet entitled "Stop *Caulerpa*!," from production of a film on the battle against *Caulerpa* at Agay to presentation of the film to the public.

But the following autumn, six months after the electoral fever, there was less concern about *Caulerpa*. During the summer of 1992, the municipal firemen had indeed torn out several small colonies discovered 300 meters north of the sheltered harbor, but they had left the *Caulerpa* that had been saved as an experiment in the small harbor. Thus, in November 1992, I was not surprised to find 50 square meters of *Caulerpa taxifolia* in the harbor. The sheet was still there, but it was now under the *Caulerpa*, which had contrived to grow around the edges of the sheet and over the plastic, and then had implanted itself in the sediment that had settled there.

Recognizing that the struggle had been abandoned, a local dive club undertook, for two successive years, clandestine eradications by manu-

ally removing colonies that began to spread in the bay at Agay. A new,   **135**
drastic method was tested: divers poured underwater concrete on the
Caulerpa. They treated in this fashion an area considerably larger than
that of the "official" eradication by the firemen. But the activities of
these clandestine volunteers did not succeed in eliminating *Caulerpa*
from Agay. At the end of 1994, 800 square meters were located there in
patches scattered over three kilometers of coastline. By the end of 1996,
the alga had invaded several dozen hectares in the bay at Agay, and it
was discovered in four inlets located more than four kilometers away
both to the east and to the west.

After the highly publicized initiative of the mayor of Saint-Raphaël,
many people latched onto the idea of a complete eradication. People
began to eye the underwater eradication site enviously. I had suggested
mobilizing the dive clubs with the logistical support of the navy and its
divers; many applicants made it clear to me that this was a matter for
the very exclusive caste of professional divers, the only ones used to
working underwater.[40] Thus it came to pass that, at the end of the spring
of 1992, the majority of the major underwater contractors had submitted proposals to IFREMER. This was also a period when the most cocka-
mamie inventions were concocted to master the killer alga.

Some inventors solicited my collaboration in their schemes, hoping
to get my support. Others hoped that I would ratify their testing approach, apply with them for a patent for their invention, or fund them
to bring their method to fruition. Respecting the official hierarchy of
the Scientific and Technical Committee, I forwarded all their requests
to the local branch of IFREMER.

ERADICATION METHODS CONSIDERED

While the private initiatives proliferated and the local community got
involved, the local IFREMER experts were content simply to log in
comments. All this was not easy to manage. It required simultaneously
feigning interest in "candidate eradicators" and showing that the experts were working on the problem. In the face of constant media pres-

136   sure, IFREMER agreed to take up the matter with funding of 150,000 francs from its own budget. But, with little equipment and few personnel suited for work on the small bottom patches, the institute joined the battle by proxy. A small private company from Bouches-du-Rhône—High Tech Environnement (HTE)—was given a contract to test the most promising techniques selected from the file. The tests were pompously designated as "validation procedures," "feasibility tests," or "prequalification operations." IFREMER was ultimately in charge of all these projects and played the role of approval office for the methods.

The proposals were classified by several themes. Chemical elimination entailed applying copper ions or salts, quicklime, chlorine, or herbicides or algicides that were already available. Physical elimination included treatment of *Caulerpa* by ultrasound, destruction by explosives, freezing by application of dry ice, scalding *Caulerpa* by covering it with a bell jar full of hot water, burning it with a blowtorch, and even a "bioelectronic" treatment. Manual removal was also contemplated; diving "Schwarzeneggers" did well to remove ten square meters an hour by hand. Burins, trowels, hammers, and hatchets were also tested in an attempt to accelerate the harvest. An Italian colleague suggested in all seriousness harrows fitted with huge scythes that could have been designed by Leonardo da Vinci. Agricultural equipment adapted to the seas was solemnly presented as the final solution. Other, more credible approaches were tested that allowed more rapid tearing out and harvesting of the algae, such as aspirating the *Caulerpa* or using a high-pressure water jet to dislodge the thalli. Biological control was also suggested. I received pressing appeals to introduce the dugong (a marine mammal) into the Mediterranean. One researcher announced to the press the imminent perfection of a biological control technique using the sea hare, a huge slug known to devour Mediterranean algae (although not the native *Caulerpa* species of the Mediterranean).

From June 1 through 5, 1992, five months after having been assigned the problem of eliminating the alga, IFREMER and its contractor HTE tested the methods selected from the many in the file. The firms chosen to participate in the "Grand Eradication Competition" for

Caulerpa taxifolia had to make their methods operational. The work was part of their demonstration of their know-how and thus was at their expense. The hope of obtaining eventual contracts was their sole motivation. Some firms had confidence in their ideas and mobilized heavy equipment: barges and trucks. They had to keep an expense log, giving the place, the area treated, and how long it took. The region around Cap Martin had been chosen for the demonstrations. Fishing and anchoring were forbidden at the site during the exercise,[41] and markers delimited the theater of operations. The mission assigned to HTE was to estimate the cost and efficiency of each method.

The situation was amusing. Tough men, experienced in underwater construction, used to soldering or connecting pipes and chains underwater, had to rip out the delicate green fronds without dispersing bits and pieces. Each worksite was monitored by a diver-judge who filmed everything. The professional divers, expecting to remove the green algae rapidly, discovered underwater what *Caulerpa* was really like. Results did not live up to expectations. No firm subsequently got an eradication contract. On the other hand, these tests allowed the government experts to convince themselves of the difficulty of the enterprise.[42] A first estimate of the cost of eradication was drawn up:[43] ten to twenty square meters could be cleaned in an hour by a team of two divers for a cost of 30,000 to 50,000 francs per day (with a maximum of three dive-hours per diver).

The evaluation of the validation tests was to be presented at the third meeting of the Coordinating Committee on *Caulerpa taxifolia* on June 22, 1992 at Marseilles. We watched films of each experimental method, after which there was a byzantine debate on the advantages and disadvantages of the different methods or the technical improvements to incorporate. This discussion would only make sense if it was immediately followed by action. But on this point, there was never any debate—inertia ruled. The thirty or so assembled experts and high officials, some of whom had come from far away, enjoyed hours of concocting miracle techniques, with no thought to when they would be applied or how they would be financed.

138   I restated for the benefit of the representatives of the Ministry of
the Environment and the Secretariat of State for the Sea that the whole
matter was essentially one of political will. Either we immediately engage in the battle and use the best methods known to us, or we could
wait. But, in that case, it is important to consider that the thirty invaded
hectares, highly circumscribed now, could rapidly grow to three hundred
hectares with less well-defined boundaries and at depths greater than
twenty meters, thus becoming much more difficult to handle. To the
cost of the eradication should be added that of searching for isolated,
scattered cuttings around each site and, finally, that of returning several
times to each site to remove possible regrowth. The cost would surely
grow by a factor of ten or twenty. The only other option was to give up,
surrender to the killer, but then we would need to have the nerve to say
this and to orient our research differently.

Once again, I emphasized the contradictions in the official stance:
to justify the decision to intervene, the ministers wanted to know
more;[44] at this time, the end of June 1992, no laboratory had received a
single cent from the national government to undertake the research
they wanted to see. In fact, it seemed obvious that the ranking officials
in Paris were not convinced that *Caulerpa taxifolia* presented a substantial risk and, a fortiori, that eradication was warranted. So why should
they finance research that would not confirm their opinion? The tests
of eradication procedures had not cost the national government much
and allowed it to procrastinate.

Faced with so much contempt for the gravity of the situation, so
much incoherence, and so much ignorance, I asked the representative
of the Secretariat of State for the Sea to define the official government
position as soon as possible. The co-chairs of the Scientific and Technical
Committee also hoped to receive the minutes of the meetings on *Caulerpa taxifolia* that took place behind closed doors in Paris, under the
aegis of the Interministerial Committee for the Sea. We knew that the
government handling of the affair was defined during these meetings.
The decision-makers had, as their only advisors, IFREMER experts who
presented, with their own questionable interpretation, the scientific as-

pects of the problem. During the meeting, the representative of the Secretariat of State for the Sea promised to send me these documents, all
the while warning me that they were highly critical of our hypotheses.
As I wanted to know the basis for the views of the interministerial committee, she told me that letters and reports of specialists and of organizations had been examined in high places and that their opinions were
contrary to ours. The conversation took place in a corridor at the end of
the day; this was how a high official, in a hurry to finish and return to
Paris, disposed of the essence of the subject.

I never received the minutes in question. By contrast, the representative of the national government (then run by the Left) officially sent
me, for my own information, one of the documents that had been referred to: a defamatory letter received by her agency through the inter-
mediary of the vice-chair of the Coordinating Committee on *Caulerpa
taxifolia*, who was also the president of the marine commission of the
Regional Council of Provence–Alpes–Côte d'Azur (which had a majority from the Right). This letter, one of those that had "captured the
attention" of the Interministerial Committee for the Sea, had been written by the president of an association in Var that militated for underwater reestablishment of *Posidonia oceanica* meadows using methods that
my colleague Boudouresque and I had often criticized. On four pages
about *Caulerpa*, we were depicted as mentally ill and described by other
equally laudatory adjectives. This was the kind of literature that the
ministerial strategies were based on.

SOME TIMELY ERADICATION OPERATIONS

Having received European financing for the eradication, IFREMER, assisted by several divers from the navy and the firm High Tech Environnement, was motivated to proceed with several attempts at eradication. These operations were all conducted in a contractual framework.
IFREMER used navy divers for a set number of days and had paid HTE
for a specified number of hours of underwater work. Instead of using
this manpower to attack the sites where eradication could have suc-

ceeded, the IFREMER technicians chose two sites where the area colonized by *Caulerpa* greatly exceeded the area that could have been treated with the means available. The divers tore out the alga by hand or with the aid of a suction device that produced water jets under pressure to dislodge the alga. But the four successive operations all remained unfinished; they did not vanquish *Caulerpa* at Toulon and Hyères.

Among these misguided operations, the one at Hyères demonstrated the problems with a hasty announcement of results. A Parisian tourist telephoned the laboratory to tell us about a two-square-meter colony off the Potinière beach opposite the large yacht harbors at Hyères. Confirmed by the local dive club, this observation spurred us to send a diver to map the colony. A report was sent to Maritime Affairs and to IFREMER. We emphasized the need to conduct a larger inspection of the surrounding bottom patches. IFREMER undertook an eradication of the small colony with local divers and the harbor authorities of Hyères. "No more killer alga at Hyères!" trumpeted a local newspaper; a long article copiously illustrated with photographs described the glorious operation.[45] But the next day, an agent of the Port-Cros National Park found 500 square meters of the alga just 100 meters west of the eradication site. Our divers once again moved to control the invasion and to evaluate the affected area. The participants in the original eradication operation restricted themselves to diffusing the bad news to the media.

Another unfortunate operation unfolded very far from Monaco, in the Department of Pyrénées-Orientales, at Saint-Cyprien near Perpignan. Discovered in December 1991, a colony began to worry the Spanish authorities. Michel Barnier, new Minister of the Environment, was invited by his Spanish counterpart to eradicate these *Caulerpa taxifolia* proliferating not far from the shores of the Costa Brava. They had not all been eliminated by a diver from the harbor and the local authorities had not taken the trouble to verify the work. Two years later, a diver from my laboratory was curious and dove there: several hundred square meters had grown. IFREMER was assigned the task of preparing an estimate for the cost of treating the area of *Caulerpa*, evaluated in the spring

of 1994 as 800 square meters. The national government and the local **141**
communities tried to foist off on one another the financing of this operation. Three months later, 1,250 square meters covered the bottom of the
harbor, and the mounting bill for the eradication of *Caulerpa* at Saint-
Cyprien discouraged the decision-makers. Thus, the alga continued to
grow in the Saint-Cyprien harbor and, if nothing were to stop it, it
would escape from this site.

A TIME FOR INVENTORS

If the authorities were uninterested in the ecological problem posed by
the alga, men of good will individually devoted much time and money
to develop machines, to test methods, and even to patent their inventions.

It was thus that a Paris engineer on vacation at Menton, convinced
that steam was the best way to defeat *Caulerpa,* asked me to test his
invention, patented under the *nom de guerre* of "Caulerpa Killer." During a mapping dive at Cap d'Ail in March 1993, we undertook a test. On
a small outboard, he had installed an enormous pressure cooker on a gas
stove; attached to this pressure cooker was a long insulated pipe at the
end of which was a spray gun. Armed with this sort of steam machine
gun, I dove into water that was glacial for the Mediterranean: 12–13°C.
Taking careful precautions (a double pair of gloves), the killer of the
killer alga aimed at a pile of *Caulerpa* a meter away. I pressed the trigger: a stream of bubbles escaped from the cannon. Too far! I brought
the war machine closer to the alga and again pressed the trigger: nothing but an insignificant stream of hot water and tiny bubbles rising towards the surface. A beautiful sea anemone was able to bathe in warm
water!

I tactfully informed the Don Quixote of the killer alga that his
weapon needed some improvements. After all, it was possible that an
old steamship that could have been equipped with a steam cannon could
have produced more heat than his pressure cooker. The inventor really
wanted to contribute to the battle against the alga. He persisted in this

142   course of action and invented the "Caulerpa Killer II," using ultrasound, which was tested with little success by the director of the aquaria at Marineland in Antibes. A second procedure, also based on the use of ultrasound, was presented by the management of naval construction in Saint-Tropez (an arm of the Ministry of Defense) and tested by IFREMER.[46]

In March 1992 new hopes arose for several weeks, based on the sea hare. This large mollusc (some specimens weigh one kilogram), common in the Mediterranean, was said to devour all *Caulerpa* rapidly. A photo of the beast appeared prominently in the press.[47] The animal was even ceremonially presented to the mayor of Marseilles.[48] But the scientist who promoted this idea had not tested it beforehand. The sea hare or aplysia *(Aplysia punctata)*, partial to local algae, was not known to eat the Mediterranean *Caulerpa prolifera*. The scientist restricted it to a diet based on *Caulerpa taxifolia*. As one might have expected, the animal was hardly enthusiastic about the toxic alga, some animals becoming bizarre, stiff corpses after eating it. No one talked about the failure of the first experiments.

Another eradication method was the subject of an animated debate: the use of chemicals. It had been suggested since 1991 by Boudouresque,[49] then suggested by the local IFREMER representative.[50] But on January 24, 1992, at the meeting chaired by the Secretary of State for the Sea at Toulon, Merckelbagh (the director of the Office of the Environment of IFREMER) had expressed his categorical opposition; there would be no use of quicklime, copper salts, or other herbicides known as effective treatments of growths of freshwater algae.[51] According to him, it was not worth taking the risk of polluting the sea to eliminate what he considered a negligible problem.

Use of chemicals frightened people. There was fear that it would cause more damage than the alga had. In fact, it could have been effective and easy to use: by targeting just the invaded areas, collateral damage could have been limited. Despite the repeated requests by my colleague from Marseilles,[52] no test with chemicals was permitted by the IFREMER experts. The order, emanating from the director of that institute, was well heeded. Several months later, an engineer from Nice wanted to have copper (a well-known algicide) tested; he was politely dismissed by IFREMER.

The proposals drawn up by the directors of a quicklime factory were also brushed aside. This firm had sent me a kilogram of quicklime for an experiment. Quicklime, or calcium oxide, is not toxic in the environment. During a dive, I wanted to see its effects. The lime granules should theoretically have burned the *Caulerpa* during a strong reaction upon contact with water. I stationed myself six meters deep. My signal was relayed by a swimmer on the surface and resulted in the quicklime being tossed over the *Caulerpa* bed. The result was negative: the lime reacted with the seawater before reaching the bottom. It would be necessary to coat the granules with a retardant.

While the invasion proceeded toward irreversibility, the Monacan authorities called me. I had suggested using a freighter to pour underwater herbicides or tons of sand on the two zones that were then heavily contaminated, Cap Martin and the Monaco cape. My caller must have taken me for a fool; the idea must have raised smiles at the Prince's palace. But I was always convinced that, if we did not succeed in eradicating the alga in the coming months, the damage caused by the alga would be far greater than several kilometers of coast burned by algicides or temporarily buried under sand. Submarine life would quickly reappear in the latter circumstances.

In the middle of the summer of 1993, the mayor of Saint-Cyr-les-Lecques, who had distinguished himself a year earlier by denouncing the pseudo "ecological crime,"[53] turned to the press once again on the subject of *Caulerpa taxifolia*. Elected to the Regional Council of Provence–Alpes–Côte d'Azur and named chair of their environmental committee, he promoted an eradication method developed by two of his citizens. A retired engineer and an aquarium merchant had just produced a machine capable of destroying *Caulerpa*.[54] It used copper, which is well known for its algicidal properties. The underlying principle consisted of diffusing copper ions between two electrodes around *Caulerpa*. The amount of copper released to the environment by this method is minimal. A prototype tested in an aquarium had shown the method to be effective. Forced to consider the discovery supported by the Regional Council, IFREMER tested the method. The influence of politicians and the media was more persuasive than that of their own scientific advisors. An IFREMER expert, assigned to supervise the eradication trials, attended the first experiment on destroying *Caulerpa* in the sea with copper.

The citizens from Lecques had constructed an impressive machine. A gadget powered by batteries attached to a raft was hauled to the *Caulerpa* beds; it had two metallic disks, between which an electric field diffused copper ions. Subjected to this toxic field, the *Caulerpa* were supposed to die. In fact, several days later, they had died in the area traversed by the machine (approximately one square meter). The operation was well publicized. It was mid-August 1993, vacation time in France and a time of the year when the public is very alert to everything that has to do with the sea. The experiment was presented as the beginning of the end; the problem of the killer alga was going to be definitively resolved. The amateur inventors, hailed as heroes, had vanquished the marine monster, where scientists and government experts had failed miserably! They were honored by the regional, national, and international press, by television and radio.[55] A presentation of the method was organized before the members of the Regional Council under the aegis of the mayor of Saint-Cyr-les-Lecques.[56]

But the machine had not been perfected, and it was difficult to drag it over all the kinds of bottom patches. It left in its wake *Caulerpa* fragments that quickly regrew. The alga quickly recolonized the "cleared" spaces. Dazzled by the media attention, the inventors from Lecques threw themselves into designing a second-generation machine. From the wheelbarrow version (dubbed CEV 1 by its designers),[57] they moved to the tank model: a string of several series of copper disks diffusing copper ions (the CEV 2). The days of *Caulerpa* were numbered. But despite the repeated appeals from the inventors and their political supporters, the invention did not receive the expected reception from IFREMER. The plans for the anti-*Caulerpa* tank remained in their boxes; the wheelbarrow version was never used. Because the invention was of no further interest to the media, it was quickly forgotten. Stubborn, and always supported by their mayor, the inventors nevertheless continued to improve their system and proposed a whole line of CEV machines to be used in bottom patches of different configurations.

The following summer (1994), two other chemical eradication techniques took the stage. The first was not well publicized;[58] it was a proposal by a private firm based on the application of bleach.[59] If the chlorine had effectively killed the killer alga, it would have been necessary to find a way to concentrate this algicide on the algae. The second technique carried the prestigious imprimatur of the National Center for Scientific Research (CNRS). A chemistry laboratory specializing in permeable membranes[60] had developed a sort of cover that liberated small quantities of copper salts. The CNRS team at Montpellier[61] tested its process at Toulon in the presence of IFREMER experts. In fact, the permeable cover (called "killer cover" by several newspapers) did kill the *Caulerpa* beneath it. This technique also had its moment of glory.[62] Once again, hope was vested in this discovery much as it had been placed in the invention of the unfortunate competitors from Lecques. To be sure, these latter individuals reacted strongly, laying claim to the idea of using copper,[63] but this response did not interest the media. By the end of 1996, neither the wheelbarrow nor the cover nor other chemical methods were used on a large scale.

Far from being discouraged by the fact that the previous discoveries had not been applied, potential eradicators continued to emerge from the depths. In January 1996, the Midi Saltworks Company entered the "Grand Eradication Contest." When one has mountains of salt painfully extracted from the sea, it is easy to dream of using them for some good cause. Thus they proposed dumping the salt into the sea to kill *Caulerpa*. The chief of the aquaria at Antibes Marineland took charge of the operation. In fact, twenty kilograms of salt deposited on a square meter of algae did kill it. Subsequently dissolved in seawater, the salt had no harmful effects on the habitat. Tested in the field in December 1995 on ten square meters, the process was patented, then disclosed to the pub-

146   lic.[64] In July 1996, a large-scale operation was undertaken in the harbor of Saint-Cyprien (Pyrénées-Orientales). Ten tons of salt were poured by divers over a 400 square meter lawn of *Caulerpa*. At the end of September 1996, our team was supposed to map the spread of the alga in the harbor. With a representative of the Midi Saltworks Company present, we determined that the salt had only made a "hole" of about 100 square meters. Very dense *Caulerpa* beds surrounded this hole. This *n*th attempted method made everyone smile, and the wits did not pass up the occasion to make it a clever joke: just a little pepper and oil and the salade Niçoise will be ready![65]

A NEGATIVE BALANCE SHEET

All the proposed methods destroyed more or fewer algae and had been tested on small areas. The effectiveness of their application to large, heterogeneous zones, with the goal of destroying every *Caulerpa*, was completely untested. All required divers to do the work. It is the need for diving that is their most onerous feature. The hourly cost of professional divers is very high and their total daily dive time is carefully regulated for safety purposes.

It was thus necessary to increase the speed of the eradication. But, whatever the method, we could aspire to lower the cost of the eradication only by a factor of ten; one diver using one of the systems could clear at most an area ten times larger than that which could be cleared by ripping out the algae by hand. The best price for the destruction of the alga was estimated at 200 to 400 francs per square meter. The algal cover increased in some sites by a factor of ten each year. The improved rate of clearing allowed by the various eradication techniques relative to the rate of algal growth was laughable.

Moreover, zones had to be completely denuded to prevent the alga from simply resprouting. To the cost of the eradication operation was thus added the cost of detailed inspection of the treated section, which required many dive hours, especially as it would be necessary to return to the site repeatedly to verify that cuttings had not produced new growth. The cost of these various necessary procedures is, of course, pro- **147** portional to the area invaded and is independent of the eradication technique employed. All these factors argued for immediate action, whatever the technique, beginning with the small isolated colonies at Saint-Cyprien, Toulon, Hyères, Le Lavandou, and Agay. However, no such effort was mounted by IFREMER, whose sole mission was to oversee the development of eradication techniques, not to apply them.

At the end of 1992, IFREMER lost control of the project; in the absence of a decision, the "eradicators" were freed from supervision and competed instead with one another. The teams whose performance was assessed by IFREMER always announced, via the media, that they had the definitive solution. Rejecting inaction, they were somewhat shocked that no one called on the procedure that they had laboriously developed at their own expense. But the efficacy of the various methods on large, heterogeneous areas was far from proven. Here and there, anonymously, small colonies were nevertheless eradicated successfully by biologists and independent divers who simply pulled out the algae by hand.

It was in the waters of the Port-Cros National Park that, for the first time in France, an officially sanctioned battle was mounted against the alga. In the autumn of 1994, bottom patches in the park were inspected by dozens of volunteer divers. A colony of three square meters was discovered in the main anchoring area. After a quick interaction among the park managers, they determined to eliminate the alga. It was to be eradicated by hand by four divers in one hour. Two years later (late 1996), several algal colonies were found near the zone from which the alga had previously been eradicated. They were again removed. How long can we struggle in this way to preserve this sanctuary for Mediterranean ecosystems?

The forty-odd eradication operations inventoried for *Caulerpa taxifolia* constituted a considerable amount of work. If IFREMER used 500,000 francs studying these techniques, a million francs can be estimated as the total sum swallowed up in France by the entire gamut of "eradicators" in research to perfect a miracle method. Despite these efforts, they have all observed the limited scope of their actions. All have recognized that their contribution to the battle was modest, all the more so because, if the alga has reached a site once, it can do so again. They have especially noted bitterly that what is most lacking is a global battle plan, a comprehensive scheme drawn up through intense interaction among the affected countries.

The Elected Officials and Their Experts

The involvement of elected officials, administrators, and experts was not equally intensive and, overall, was very disorganized. The change in government in March 1992 did not materially change the story.

THE MINISTERIAL MERRY-GO-ROUND

In mid-summer 1992, there was much excited talk about *Caulerpa taxifolia* in the press and on television. In quick succession, two ministers appeared on the highly publicized scene developing in the South of France. Ségolène Royal had just succeeded Brice Lalonde as Minister of the Environment when the alga was found in the harbor at Villefranche, several kilometers from Nice. The assistant mayor of Villefranche-sur-Mer, in charge of the environment, asked my advice on eradication. I explained that this operation was supported by agencies of the national government. It was necessary to tear out the alga and to organize follow-up work at the site to remove possible regrowth. This was a contractual sort of work that I could not undertake as part of my university functions and service on the *Caulerpa* committees. He therefore made a request to the Ministry of the Environment, which agreed to sponsor the operation under the direction of the minister in person. A local dive club was pressed into service to attempt an eradication with the assistance of an underwater vacuum cleaner. The killer alga was torn up, vacuumed, collected, and a clump was presented to the minister, to the applause of the prefect and an august assemblage of deputies, senators, and mayors of coastal regions, all crammed into a fragile floating wharf. The *Caul-*

erpa "bouquet" offered to Ségolène Royal was immortalized by a number of newspapers. The scientists gathered for the occasion presented
the results of their research. They made it clear that they were progressing slowly without funding. The minister promised to see to the
disbursement of the funds that had been promised, authorized, and
blocked by the agencies under her predecessor.

The results were completely predictable. At Villefranche-sur-Mer,
no government agency had been assigned surveillance of the site and
elimination of regrowth. Six months later, under the floating wharf that
had been the theater of the ministerial eradication maneuvers, my divers discovered the alga covering twenty square meters. The authorities
were immediately informed—and the alga was allowed to grow. By the
end of 1994, more than 7,000 square meters of *Caulerpa taxifolia* covered the bottom of the harbor and began to spread to the entrance pass.
By the end of 1995, the majority of the harbor was invaded and the
bottom at the entrance was colonized.

A week after this memorable visit, Charles Josselin, Secretary of
State for the Sea (successor to Le Drian), came to Hyères accompanied
by Paris reporters. During the press conference, he read a misleading
statement on the origin of the alga and the eradication: "Charles Josselin noted with satisfaction that the small patches observed in various
parts of the Mediterranean arose from local releases and not by spreading from either of the two principal infestations (Cap Martin and Monaco). These small patches are easily and systematically eradicated, either by navy divers or by agencies of the affected municipalities in
concert with the Secretariat of State for the Sea."[66] Contrary to this deceptive declaration, the patches of *Caulerpa* had not been eradicated. At
Toulon, at Hyères, at Lavandou, and at Saint-Raphaël–Agay, the areas
invaded by the alga had doubled or tripled in one year. By autumn of
1992, the total invaded area reached more than 400 hectares. In June of
that year, the alga had appeared in Italy and Spain.

A third ministerial visit took place in October 1992 on the occasion
of the fourth meeting of the Coordinating Committee, held in Toulon.
Beforehand, the Scientific and Technical Committee had analyzed the

150   scant knowledge acquired without government financial support and had admitted, in my absence, the existence of an environmental risk, recognizing what I had been continually saying: the *Caulerpa* invasion had reached the point beyond which it was too late to envisage a total eradication. They asked the Coordinating Committee to "control" the invasion by eradicating colonies isolated from the most heavily invaded area on either side of Monaco–Cap Martin. Josselin came again, this time to preside over the Coordinating Committee discussion. The meeting was especially well prepared. During the press conference, the reassuring declarations and an optimistic summary of the situation had been edited and printed in advance by the office of the Secretary of State for the Sea. The message rested on the single observation that the scientists had been unable to prove that the *Caulerpa taxifolia* toxins accumulated in fishes. The alga was thus not toxic for humans. The entire affair was thus traced to one anecdote, the importance of which was artificially inflated by the scientists.

During the meeting, the scientists presented their first results on the effects of *Caulerpa taxifolia* toxins on biological models. The observed level of toxicity supported the classification of the caulerpenyne molecule as moderately toxic. The problem lay in the elevated concentration in the alga. From July through January, the biomass of *Caulerpa* in the field was substantial.[67] The specialists emphasized the fact that there was little knowledge of the fate of the toxins, which changed quickly into products that were less well known but that could have persistent toxic effects. Knowing the threat of unlimited spread of the alga, they called for the urgent eradication of colonies outside the maximally infested zone. This principle was accepted by the Secretary of State for the Sea. But no timetable was drawn up to execute this action, and no budget was established for it.

The Scientific and Technical Committee had neglected one factor in the new battle plan: spread of *Caulerpa* by boat anchors. It accomplished nothing to eradicate the small, isolated colonies if one did not block this very efficient means of dissemination. Anchorage in areas invaded by *Caulerpa* (which at the time constituted less than ten kilometers of coast) had to be prohibited. I raised this issue, which had not been on the agenda, and a lively discussion ensued. Admiral Tripier, Maritime Prefect for the Mediterranean, advised installing mooring buoys in the invaded zones, which would allow boats to use the contaminated sites without dropping their anchors into the algae on the bottom. His proposal was accepted by the committee and reported by the media but was never put into effect.[68]

The change of ministers had thus caused no change in the strategy of the national government, which one could sum up by the slogan "let the situation rot." Regularly summoned before the committees to present their work, the alarmist researchers were kept under control. One hoped that, in the absence of research funding, they would become discouraged. There was even surprise in high places at the resistance of scientists to this "torture" that was being inflicted: "At the Ministry of the Environment, it was moreover stated that it was not known where the laboratories got their funds to work on the alga . . . but the reports were expected!"[69] The main objective was to reduce what was, for the national government, the sole problem: the media catastrophism. In his press conference of July 29, 1992, the Secretary of State for the Sea regretted "the alarmist declarations, without a real scientific basis, publicized recently in the *Caulerpa taxifolia* affair."

The director of IFREMER raised the stakes in the *Caulerpa* affair by vitriolic phrasing in a long article crowing about the merits of the institute in its assessment mission: "the alarmist statements issued by certain university scientists about *Caulerpa taxifolia* in the Mediterranean demonstrate very well the dangers of behavior that discredits their expertise in the long term."[70] The director of the environment and coastal management of IFREMER must have revealed the underlying reasoning of the official stance in an internal institute bulletin.[71] In an article entitled "*Caulerpa taxifolia:* myth or reality?"—in which he adopted an attitude of reassurance—he affirmed what was the essential point in his eyes: "the only known damage to date is that caused by the media." A local expert, press secretary for IFREMER, produced a more original version: "They're making a 'noir' novel out of this, a whodunit thriller that's causing a bit of harm. There is no real danger, so there's time to talk about it." [72]

Fortunately, we had some research funding that had been granted to us by local communities before the establishment of the Coordinating Committee for the *Caulerpa* affair. The allocated sums allowed us to continue for a year. Also, our laboratory, associated with an INSERM team, was able to begin studies on the toxins by 1992 and to organize a second publicity campaign targeting dive clubs on the French Mediterranean coast. At the end of 1992, we were able to provide an update on the status of the invasion. But for want of specific, dedicated funding, other avenues of research saw no progress. In the laboratories concerned with the matter, we became experts in administrative semantics and the nuanced differences between funds released, allocated, announced, and available. We also evolved into experts on the administrative notion of a time scale; the ministries exacerbated the problem by employing quarterly, biennial, or even annual disbursement schedules. We lived with all manner of official annoyances, from delays at all stages in drawing up a file on the problem to the ad hoc slicing of sums promised by the Secretary of State for the Sea (200,000 francs rather than the 500,000 promised in January 1992, distributed among eight payments from December 1992 to 1994), not to mention the occasional "misplacement" of our file concerning the sums promised by the Ministry of the Environment (500,000 francs). Often, we had to pay bills ourselves when future reimbursement was uncertain or deferred. An examination of the many unkept promises is edifying. Of the total sum designated as released, only a small fraction was actually disbursed. We were hostages to a pernicious system, where political strategies dictated that scientists spend large sums for their research, but that appropriate decisions would be made only after the results were in.

The evident ill will and lack of fulfillment of the promises made by the two ministers generated a lively debate with my colleague Boudouresque. I was again inclined to abandon all participation in the committees on *Caulerpa taxifolia*, while nevertheless trying to pursue the research, but without disclosing how much progress had been made while still waiting for the promised funding. Boudouresque was more confident in the government authorities and more conciliatory. I reluctantly followed his proposed course of action. My room for maneuvering was too small to do otherwise. I did not want to oppose my colleagues who, by and large, wanted to base their opinions on solid research.

A TURNAROUND: THE EUROPEAN UNION GETS INVOLVED

Without financial support, the scientific research could scarcely proceed, and the assessment of the problem came to a standstill. Boudouresque did not skimp in trying to negotiate with all potential sponsors. The key assistance was to come from the European Union; the "green" deputies of the European parliament pressed for E. U. financial support to deal with the *Caulerpa taxifolia* problem. Helped by Boudouresque, they drafted a manifesto on the subject.[73] From that point, the French Ministry of the Environment agreed to involve itself, promising its support to bring to fruition a European research program. The project was also strongly supported by an unexpected ally: the Italian organization Castalia (with bases in Rome, Genoa, and Naples), which specialized in impact studies and was supported by a powerful lobby in Brussels (seat of the European Union).

All these pressures would doubtless never have succeeded but for a fortunate happenstance. As the end of the 1992 budget process neared, the environmental directorate had unexpected funds on hand. The construction start of a European center for management of the environment, scheduled for 1992, had been canceled at the last minute over disagreement about the choice of host country. The funds reserved for this construction were available, and it was absolutely necessary to commit them before the end of 1992. A delegation of ten scientists convened in Brussels, under the direction of Boudouresque, to refine the hastily drawn up research program. It was an ambitious program: one million E.E.C. ecus were to be shared over two years among more than twenty French, Spanish, and Italian laboratories. The prospect of substantial funding led many people to revise their opinion of *Caulerpa*.

In France, the end of 1992 was marked by a new legislative election campaign that led in 1993 to the collapse of the Socialist majority. The campaign was an occasion for the government to congratulate the various protagonists in the *Caulerpa taxifolia* affair. Even the most alarmist scientists received letters of congratulations from the Secretary of State for the Sea thanking them for their research and their dedication in coming to grips with this problem. In the chorus of praise, Doumenge was promoted to Officer of the Legion of Honor.[74] During this time, the funds promised by the Socialist ministers arrived in dribs and drabs. At the end of 1992, while all funding was stopped up at the Ministry of the Environment, the Secretariat of State for the Sea disbursed a total sum of 20,000 francs.

THE GOVERNMENT CHANGES . . . AND DOES NOTHING

In March 1993, the Right succeeded the Left in France. With this political change, the Secretariat of State for the Sea disappeared. Matters relating to the sea were transferred to the Ministry of Agriculture, while the Ministry of the Environment essentially took charge of the marine environment. One year after the designation of Michel Barnier to head that ministry, no manifestation of interest in *Caulerpa* could be observed. None of the *Caulerpa* committees, created with a tenure of two years in February 1992, had been renewed. The fifth and last meeting of the Coordinating Committee, held in November 1993, confirmed the flagrant lack of interest on the part of the national government. The financial shortfalls and the unkept promises were key subjects of discussion. At the end of the meeting, the interregional director of Maritime Affairs thanked the committee members, affirming that the work should be pursued in the form of a "national research program." The Scientific and Technical Committee on *Caulerpa taxifolia* was to be transformed into an interagency committee charged with communicating the fruits of our labors to the Coordinating Committee for Marine Research and Technology Programs. Based on this promise of metamorphosis into shadowy or evanescent committees, the *Caulerpa* committees disappeared. Subsequently, neither the national program nor the phantom committees were activated. In January 1994, no ministerial representative deigned to travel to Nice to attend the first international colloquium on *Caulerpa taxifolia* organized by the scientists involved in the European research program. The media pressure had fallen. At this time, not one single centime of the 500,000 francs destined for research on *Caulerpa taxifolia*, and promised successively by Lalonde and Royal, had reached the scientists, despite many reminders of duly signed agreements. Worse, the withholding of research credits blocked the final disbursement of funds granted by the European Union. This situation was all the more surprising given that the new Minister of the Environment, Barnier, had written a book on natural catastrophes in which he mentioned the dangers of the *Caulerpa taxifolia* invasion.[75]

In February 1994, I succeeded in attracting the attention of a local political personality with influence on the new government,[76] who asked the minister to examine the *Caulerpa* file. The same month, Commander Cousteau shared his anxiety (by letter) with Barnier. He had sent an observer to the colloquium on *Caulerpa taxifolia*, and this person had reported the bad news to him: the alga was spreading and was seriously threatening the coastal marine environment. Thus Cousteau established for the record that, at this time, he was squarely in the alarmist camp. A new article appeared in the journal of his foundation on the risks associated with the invasion of *Caulerpa taxifolia*.[77] Finally, the minister was also called by his Spanish counterpart, concerned to see the alga proliferate on the French Catalan coast and not wanting to see it reach the Costa Brava.

Beseeched by all sides, Barnier assigned a member of his cabinet to make a circuit of the scientists involved in the research. The minister wanted to meet in Paris with the principal players in this affair so that they could describe the results of their studies in the presence of reporters. This event was the first in a series of meetings on the environment, dubbed "Ségur discussions" (by analogy with the famous "Bichat discussions" that saw the regular gathering of major medical specialists at the Bichat Hospital). There was another goal: the bureaus of the ministry, which until then had been dispersed in a Paris suburb (Neuilly) and Paris proper (Boulevard Saint-Germain), had just been reorganized into a single location, situated in a prestigious Paris neighborhood, Ségur Avenue, not far from the major ministries and a short walk from UNESCO. They were eager to publicize this "tour de force" of the minister. The creation of the Ségur discussions was a good way to promote the new, prestigious trademark, "Ségur Avenue—Ministry of the Environment." The first discussion theme chosen had to be one of the most publicized; *Caulerpa taxifolia* was selected to inaugurate the discussions.

Whatever the reasons for this new and urgent request, it was an opportunity for us to recall the unkept promises of the previous ministers. We agreed to come to Paris only after having received formal assurances that a sum would be disbursed corresponding to that agreed on in the conventions signed in 1992. The meeting, held on March 21, 1994, also had the advantage of attracting the minister's attention to the subject. He lunched with the scientists and listened to their fantastic story of the alga and the litany of pitfalls encountered from the outset. Sensitized to the matter, he named an official ministry representative to deal with the problem of *Caulerpa taxifolia*. As the "*Caulerpa* czar," he designated André Manche, a man well known on the Côte d'Azur for having managed the Port-Cros National Park very well for ten years, with spirit and enthusiasm.

Manche's drive to assess the *Caulerpa* file slowly converted the undecided decision-makers, who until then were the majority in the Ministry of the Environment and the former Secretariat of State for the Sea (which had become the new Ministry of Agriculture and the Sea). He took his charge seriously enough to dive himself at the majority of invaded sites to see the situation firsthand. He gathered the numerous scientific publications, which all demonstrated the existence of a major environmental threat. Quickly convinced, he had to spend a lot of time explaining the situation to ministry bureaucrats. The only ones who remained unshaken were those responsible for (but not guilty of) the four previous years of inaction and blocked funding. In January 1995, he succeeded in freeing up the funds that had been promised to the scientists since the beginning of 1992, and, the same year, he convinced the minister to support the researchers' proposal for a second European research program on *Caulerpa taxifolia.*

With the election of Jacques Chirac as President of the Republic, another ministerial reshuffling occurred during the summer of 1995. Corinne Lepage thus became the fourth Minister of the Environment to deal with the *Caulerpa taxifolia* affair. Until the end of 1996, her ministry took no evident position on the *Caulerpa* file. Manche had his mandate renewed, but without real power; without funding, his room to maneuver was limited to his personal communication efforts.

In sum, from the beginning of the affair, whichever ministers were in charge, the national government never took a clear position and upheld it. This failure to take charge greatly slowed—indeed, stopped—the flow of useful decisions, the involvement of branches of the agencies, and the support of the regional communities.[78]

THE ELECTED OFFICIALS IN THE REGIONAL COMMUNITIES ARE DIVIDED

During the summer of 1992, Jean-Claude Gaudin, chair of the Regional Council of Provence–Alpes–Côte d'Azur, had denounced the use of the word "killer" after his agencies had received many calls expressing fear "of a danger coming to our coasts." For him, "it was the trademark of our region, which suffered because of it; we didn't need that."[79] He nevertheless thought clearly about the matter and encouraged the researchers. He even congratulated himself for having represented the first community (in the Regional Council of Alpes-Maritimes) to have revealed the scope of the problem, having agreed to fund the first campaign to publicize and to map the invasion.[80]

The attitudes of the chairs of the regional assembly committees were not as clear. Jean-Pierre Guiran, chair of the environmental committee of the Region and mayor of Saint-Cyr-les-Lecques, had seemed skeptical about this matter since the "ambush" in his town. He was mainly interested in promoting and financing the "*Caulerpa*-killer ma-

158   chine" that two of his citizens had invented. Andrée Heymonet, chair
of the marine committee of the Regional Council and vice-chair of the
Coordinating Committee on *Caulerpa taxifolia* from 1992 to 1994, was
always indecisive about *Caulerpa.* Within her committee, the representatives were divided.[81] Although in favor of the subsidies allocated by
the Region to the scientists, some representatives hoped to support less
alarmist scientific teams, while others felt that "the role of elected
officials is not to cast continual doubt on the conclusions of the scientists,
but rather, to use these conclusions to make clear choices showing political willpower."[82] Anxious to know more, Heymonet participated in
many meetings with the scientists and organized two informational
meetings for her committee. The first, held at Hyères in September,
1995, was devoted to hearing the opinion of the representative of the
Ministry of the Environment.[83] During the second, organized at the
Oceanographic Museum of Monaco on January 30, 1996, the representatives saw films on *Caulerpa taxifolia,* heard presentations by the main
defenders of *Caulerpa,* and visited the magnificent aquaria and the laboratory of the European Oceanographic Observatory. The other regional
councillors remained indifferent for the most part.[84]

Much less sensitive to the polemic, the Region of Corsica, although
not stricken by the problem, regularly supported (by means of its Office
of the Environment) publicity campaigns to prevent any arrival of the
alga in its waters.

As for the departments directly concerned with *Caulerpa,* at the beginning of the affair they had confidence in the publicity campaigns
of the divers and in the operations mapping *Caulerpa* colonies. Alpes-
Maritimes (1991 and 1992), Var (1992), and Pyrénées-Orientales (1992)
also participated, in proportion to their size, at the rate of 40,000
to 60,000 francs per department. Beginning in June 1992, Alpes-
Maritimes stopped all assistance and the representatives of the environmental office of the General Council of that department ceased being
interested in the problem of the alga there. The elected officials did not
want to take sides in the polemic between the scientists of the University of Nice–Sophia Antipolis and the *Caulerpa* enthusiasts based at the

Oceanographic Museum of Monaco. This decision broke up a pattern of regular cooperation between my laboratory and the General Council of Alpes-Maritimes. Beginning in the 1970s, we had taken part in the study of marine reserves, of artificial reefs, and of *Posidonia* meadows in the department. Our laboratory received financial assistance, while the employees of the General Council were trained in techniques of observation and mapping of marine bottoms. This abrupt cessation of joint activities showed how fragile was the coherent management of the environment. At any moment, the sudden whims of elected officials could interrupt a project that would have to be long-term to be effective.

Finally, at the municipal level, participation by the mayors who were the most concerned with the invasion had not been planned in the *Caulerpa* committees established by the national governmental agencies. They had subsequently received permission to be represented as observers. Dissatisfied at having been thus distanced from a problem that directly affected them, the mayors formed an association that aimed to know the Mediterranean flora and fauna better and to defend them more effectively.[85] The first two meetings of this association, covered by the media, were held at Menton[86] and at Villefranche-sur-Mer and were devoted to *Caulerpa taxifolia*.

Overall, it must be said that no elected official attended closely in the long term to the algal invasion problem. Their activities were more ephemeral, more concerned with media impact than with really coming to grips with the problem. It is true that, given their level of expertise, the astonishing spectacle of dissension among scientists and experts made it difficult for the local officials to study the matter and to understand what was really at stake. Some of them preferred to hold back, while others granted subsidies to the most persistent and convincing experts. Finally, some of them took the tack of ignoring the uproar and even denouncing the entire debate as futile at best and, at worst, injurious to the reputation of the Côte d'Azur. Faced with this disarray, one might have thought that the adoption by the IFREMER experts of the hypotheses of the alarmist scientists could have dissipated the inaction and mistrust. But nothing of the sort happened.

THE MOBILIZATION OF THE IFREMER EXPERTS

From 1992 to 1996, the IFREMER experts, chief advisors to the ministers, increasingly distanced themselves from the views of the Oceanographic Museum. They participated actively in the well-funded European program of research. They devoted themselves, as we have seen, to research on a method of eradication, but also to mapping *Caulerpa taxifolia*. To know the precise location and pattern of spread of *Caulerpa taxifolia* was crucial for a proper grasp of the nature of the phenomenon. These data, easily verifiable, demonstrated the explosive character of the algal proliferation on all kinds of sea bottoms.

The IFREMER experts began to verify our maps. In February 1992, the IFREMER phycologist Thomas Belsher was assigned to plot the invaded areas described in our publication in *Oceanologica Acta* and in my first report given to the authorities in December 1991. He used a video camera towed over the bottom by an IFREMER ship. He was also charged with attempting to verify Doumenge's statements that the alga grew preferentially—and spontaneously—in polluted areas. Belsher's observations confirmed our conclusions: our maps were accurate; we had even slightly underestimated the range of *Caulerpa* or its density in certain sectors. By contrast, Belsher found no *Caulerpa* near the main outfalls of urban wastewater between Toulon and Nice.

Six months later, in the autumn of 1992, IFREMER organized a new mapping project. Belsher was to confirm our descriptions of the spread of *Caulerpa*. He had use of an old, wheezing trawler, always on the verge of breaking down, that IFREMER had transformed into an oceanographic vessel: the *Roselys II*. Belsher invited me aboard to compare our maps and to discuss the biology of the alga. The ship was moored in the Nice harbor. At the same pier, by chance, was another IFREMER boat, the *Noroît*, a magnificent research vessel equipped with the most modern mapping tools (lateral sonar, a global positioning mechanism linked to a computer that recorded the least movements of the vessel). It had on board an oceanographic submarine, the *Nautile*. The *Noroît*, the jewel of French oceanography, had been chartered by a

champagne company to find Saint-Exupéry's airplane, which an archivist had believed would be found not far from Nice. It would have nothing to do with *Caulerpa*! The *Noroît* and the *Nautile* have completed
many circuits in the Baie des Anges and have never found the plane.
During the same period, the *Roselys II* towed a rudimentary tool—a
video camera mounted on a kind of sled—over the bottom. Despite the
swells and the jolts of the underwater sled, Belsher spent hours in front
of the TV monitor to keep the system working properly. The *Caulerpa*,
the *Posidonia*, the sand, or the rock continually jumped or disappeared
from the viewing field altogether as the tow cable jerked forward. I accompanied Belsher for a day; I held out for only two hours before I declared myself out of commission!

A succession of mapping missions proceeded. In December 1992,
the navy made a ship available to the researchers, the underwater exploration and work vessel *Triton*, and an observation submarine, the
Griffon. With great emotion, I participated with Belsher in this mission
to determine the lower depth limits of *Caulerpa*. Dressed in my lieuten-
ant's uniform, I served my sixth reserve stint on board the *Triton* and
my eighteenth dive on the submarine *Griffon*. I was also moved because,
while on board, we learned the vessel was going to be scrapped; it was
to be my last mission on the *Triton*!

The sea was raging and underwater visibility was bad. We covered
three kilometers at between 50 and 100 meters of depth off Cap Martin.
Fragments of *Caulerpa* littered the muddy sand between 50 and 70 meters deep. The next day, Belsher dove in the *Griffon* off Cap d'Ail, west
of Monaco. He found *Caulerpa* attached down to 99 meters! In 1994,
1995, and 1996, Belsher returned to Monaco with another, brand-new
oceanographic vessel, the catamaran *Europe*. He also found an ally in
the Monacan office of the environment; one of its chiefs was increasingly convinced that *Caulerpa* was beginning to disrupt coastal ecosystems.

Belsher's many reports and publications all described the advancement of *Caulerpa* on both sides of Monaco.[87] The local IFREMER expert in charge of the *Caulerpa* problem, Henocque, began talking very

162   differently, finally admitting the existence of a major environmental threat. During the summer of 1995, IFREMER began to argue for the defense of biodiversity, a theme that the institute illustrated by an hourglass on the cover of its news magazine. The upper chamber of the hourglass held *Posidonia* meadows and representatives of the associated native fauna, while the lower chamber held only *Caulerpa taxifolia.*[88] This symbolic image could also have been applied to the idea of opposing hypotheses. It had taken five years for IFREMER finally to open its eyes. In an article in the same issue of the magazine, Belsher and Henocque referred to the precautionary principle in the Rio de Janeiro convention on biodiversity.[89] This was a principle that the alarmist scientists had emphasized as a leitmotif since the convention was signed in June 1992.

But if the on-site experts were increasingly convinced of the importance of the problem, the directorate of the institute remained skeptical or at least indifferent. The alarm of the people in the field seemed not to have reached the administration in Paris. Moreover, our opponents were held in great esteem by the public. In 1993, several months after the new government was installed, Doumenge was approached by François Fillon, Minister for Higher Education and Research, with the goal of educating the IFREMER branches about the activities the French had undertaken to facilitate cooperation for development. In 1994, Doumenge was also part of a committee in support of the UNI (Union national inter-universitaire) list of candidates, a higher education group close to the RPR (Rassemblement pour la République [Rally for the Republic]), the party in power.[90] In October 1995, the new government named Jaubert to the National Committee on Scientific Research, a strategic post that influences the selection of people to be promoted and themes to be developed in the life sciences.[91]

The Alga Spreads in France and Other Countries

With no eradication plans in action, the alga on the coasts of Var (Agay, Lavandou, Toulon, Hyères) and Pyrénées-Orientales (Saint-Cyprien)

had spread from an area of several hundred square meters at the end of   **163**
1992 to cover several hectares two years later, and more than a hundred
hectares by the end of 1996. Without rules prohibiting boats from
throwing their anchors on *Caulerpa*, the spread accelerated; in 1993 and
1994, the alga was discovered at Messina, Elba, Cannes, Théoule, all the
way to the waters of the Port-Cros National Park. At the beginning of
1995, it was found in two yacht harbors on the Dalmatian coast of Croatia in the Adriatic Sea. In France, the alga was observed for the first
time in Cannes, then at Porquerolles and in three new sites on the coast
of the Estérel massif. In 1996, the alga spread considerably; twenty new
sites were inventoried in France, Italy, the Balearic Islands, and Croatia.
A total of 68 sites were invaded, and more than 3,000 hectares were
colonized to some extent.

THE ITALIANS LET IT GROW . . .

In February 1992 the mayor of Menton had organized a meeting on
the Italian border and had quite naturally invited the authorities of the
neighboring communities in Liguria.[92] In turn, in May 1992 the mayor
of the Italian border city of Vintimille invited French researchers to
present the problem to local authorities. Italian divers were present.
Seeing our slides of *Caulerpa*, one of them thought he had seen it somewhere. Two weeks later, we learned that the alga had been found in the
bay at Imperia, a Ligurian municipality forty kilometers east of Monaco.[93] Upon arrival there, we found that it covered about 200 square
meters. The harbormaster had first notified scientists at the University
of Genoa. Giulio Relini's team rushed to publish the discovery, but they
were quite reserved about the threats associated with its possible spread
around the site. Later, divers found a small colony of one square meter
not far from Leghorn, 200 kilometers east of Monaco. On the advice of
Francesco Cinelli of the University of Pisa, it was ripped out.

The appearance of the alga in Italy was a major news story in
our neighboring country. Reporters baptized it by two translations of
"l'algue tueuse"—"alga assassina" or "alga killer." Two camps opposed

164

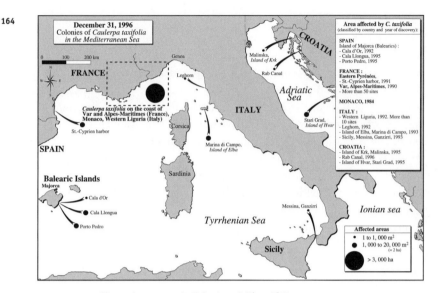

FIGURE 4.1.   December 31, 1996: Colonies of *C. taxifolia*

one another on the diagnosis of the situation: the "cautious camp," led by Relini, and the "alarmists," headed by Cinelli and Pietra, who were joined by experts of the Italian analogue of IFREMER (l'Istituto centrale per la ricerca scientifica e technologica applicata alla pesca maritima—ICRAM) and the scientists of a quasi-public society that specialized in environmental impact studies (the Castalia organization).

From the summer of 1993, a massive mobilization of dive clubs throughout Italy was undertaken: a series of articles appeared in diving journals, accompanied by magnificent photographs.[94] The main host of a televised news show on the national Italian network dove to the bottom of the harbor at Imperia to report live on the situation.[95] The powerful Italian association for the protection of the sea, Mare Vivo, distributed posters and sponsored a publicity spot on national television to encourage divers to report the presence of the alga. Brochures were distributed in all the dive clubs and Italian ports. Tens of thousands of

copies were published, under the auspices of the European program to
battle the alga, and they carried the seals of three universities (Trent,
Pisa, and Genoa), two research centers,[96] ICRAM, Castalia, and Mare
Vivo.

There were quick results from this publicity campaign. Divers reported the alga in five small Ligurian harbors spread over ten kilometers of coast both east and west of Imperia. It was quickly recognized
that this dissemination was produced inadvertently by fishermen; their
nets, trawls, dredges, or dragnets pulled up pieces of *Caulerpa* that they
tossed overboard later in the coastal harbors. It was almost certainly this
practice that brought the alga to Imperia. According to the fishermen
of Menton and the local office of Maritime Affairs of the frontier region, it was not unusual for fishermen from the Imperia region to enter neighboring French territorial waters clandestinely at night to place
their nets or to drag several trawls. They could thus have brought *Caulerpa*, which thrived just several cable lengths from the border off Cap
Martin, into Italy. During the summer of 1993, some Imperia fishermen
became angry at official inertia.[97] They were picking up more and more
Caulerpa and it was fouling their nets and trawls.

In July 1993, the alga was found in a yacht anchorage on the island
of Elba. Cinelli removed the first clumps seen, but he had to renounce
the prospect of eradicating all of it, as the alga covered more than 600
square meters. At the end of 1995, more than 3,000 square meters were
mapped at the same site. The alga was also found in July 1993 off Sicily,
thanks to the perseverance of a phycological researcher from Palermo,
Carla Fradà-Orestano. Beginning in 1992, she had inspired all the dive
clubs in her region to be attentive and vigilant. As soon as the alga was
reported to her at Ganzirri (not far from Messina), she called in rein-
forcements in the person of her colleague Cinelli.[98] They destroyed the
first three clumps that were spotted, only to notice the next day that
there were many others to the east and west.

In November 1993 she invited me to dive to examine the situation
with the help of divers from the Holothuria dive club of Ganzirri. The
alga had almost reached the Whirlpool of Scylla and Charybdis,[99] right

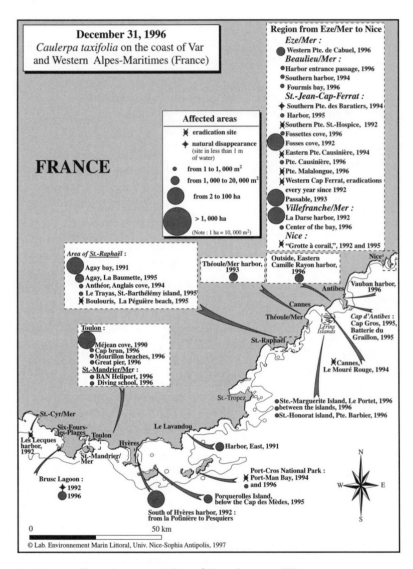

December 31, 1996

Caulerpa taxifolia on the coast of Var
and Western Alpes-Maritimes (France)

FRANCE

Affected areas

✕ eradication site

✦ natural disappearance
(site in less than 1 m
of water)

● from 1 to 1,000 m²

● from 1,000 to 20,000 m²

● from 2 to 100 ha

● > 1,000 ha

(Note : 1 ha = 10,000 m²)

Region from Eze/Mer to Nice
Eze/Mer :
● Western Pte. de Cabuel, 1996
Beaulieu/Mer :
● Harbor entrance passage, 1996
● Southern harbor, 1994
● Fourmis bay, 1996
St.-Jean-Cap-Ferrat :
✦ Southern Pte. des Baratiers, 1994
● Harbor, 1995
✕ Southern Pte. St.-Hospice, 1992
● Fossettes cove, 1996
● Fosses cove, 1992
✕ Eastern Pte. Causinière, 1994
● Pte. Causinière, 1996
✕ Pte. Malalongue, 1996
✕ Western Cap Ferrat, eradications
every year since 1992
● Passable, 1993
Villefranche/Mer :
● La Darse harbor, 1992
● Center of the bay, 1996
Nice :
✕ "Grotte à corail,", 1992 and 1995
● Outside, Eastern
Camille Rayon harbor,
1996

Area of St.-Raphaël :
● Agay bay, 1991
● Agay, La Baumette, 1995
● Anthéor, Anglais cove, 1994
● Le Trayas, St.-Barthélémy island, 1995
✕ Boulouris, La Péguière beach, 1995

Théoule/Mer harbor,
1993

Nice

Vauban harbor,
1996

Antibes

Cannes

Théoule/Mer

*Lérins
Islands*

Cap d'Antibes :
Cap Gros, 1995,
Batterie du
Graillon, 1995

Toulon :
● Méjean cove, 1990
● Cap brun, 1996
● Mourillon beaches, 1996
● Great pier, 1996
St.-Mandrier/Mer :
● BAN Heliport, 1996
● Diving school, 1996

St.-Raphaël

✕ Cannes,
Le Mouré Rouge, 1994

St.-Tropez

● Ste.-Marguerite Island, Le Portet, 1996
● between the islands, 1996
● St.-Honorat island, Pte. Barbier, 1996

St.-Cyr/Mer

Six-Fours-
les-Plages Toulon

Le Lavandou

Les Lecques
harbor,
1992

Hyères

St.-Mandrier/
Mer

● Harbor, East, 1991

N

Bruc Lagoon :
✦ 1992
● 1996

St.-Cyr/Mer

Port-Cros National Park :
✕ Port-Man Bay, 1994
● and 1996

W — E

Porquerolles Island,
below the Cap des Mèdes, 1995

South of Hyères harbor, 1992 :
from la Potinière to Pesquiers

S

0   50 km

© Lab. Environnement Marin Littoral, Univ. Nice-Sophia Antipolis, 1997

FIGURE 4.2.   December 31, 1996: *C. taxifolia* on the coast of Var

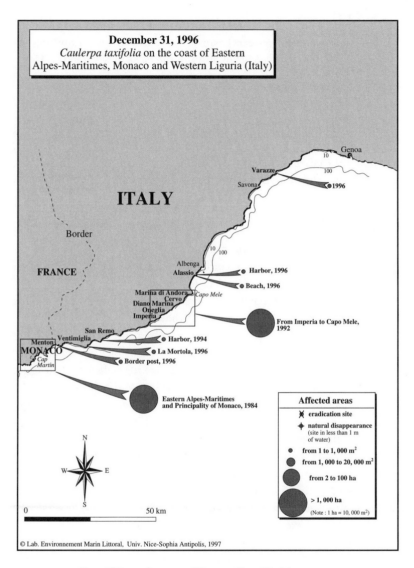

December 31, 1996
Caulerpa taxifolia on the coast of Eastern
Alpes-Maritimes, Monaco and Western Liguria (Italy)

Genoa

ITALY

Varazze
Savona
10
100
1996

Border

FRANCE

Albenga
Alassio
10
100

Harbor, 1996
Beach, 1996

Marina di Andora
Cervo
Diano Marina
Oneglia
Imperia
Capo Mele

From Imperia to Capo Mele, 1992

San Remo
Ventimiglia
Menton
MONACO
Cap Martin

Harbor, 1994
La Mortola, 1996
Border post, 1996

Eastern Alpes-Maritimes
and Principality of Monaco, 1984

Affected areas

✗ eradication site

✚ natural disappearance
(site in less than 1 m
of water)

● from 1 to 1, 000 m²

● from 1, 000 to 20, 000 m²

● from 2 to 100 ha

● > 1, 000 ha
(Note : 1 ha = 10, 000 m²)

N
W — E
S

0   50 km

© Lab. Environnement Marin Littoral, Univ. Nice-Sophia Antipolis, 1997

FIGURE 4.3.   *C. taxifolia* on the coast of Eastern Alpes-Maritimes.

168 in the strait between Sicily and Calabria, famed for its wicked currents. On one side was the western basin of the Mediterranean and on the other side the Ionian Sea, port of entry to the eastern basin of the Mediterranean. The alga was growing on the coast of the Ionian Sea. There were only small clumps of several square meters, but there were many of them and there had been no success in limiting their dissemination. To demarcate their spread, we entered the water in the little sheltered harbor where the first clumps had been discovered. Leaving the harbor, we found the current carrying us rapidly in the direction of Torre Faro, a lighthouse on the extreme northeastern point of Sicily. This was a strong, regular current; we had the feeling of swimming in a turbulent river. We inspected thousands of meters of small patches of bottom while diving in less than ten meters of water. The whole effort consisted in mastering the currents, in aiming our flippers like rudders. If we wanted to stop in order to record the characteristics of an observed clump on a plastic slate, we had to let ourselves drift or grab an outcropping. In fact, we were not at all safe; upward and downward currents also arose. After an hour, we emerged exhausted.

The Messina divers then made us dive in a sort of stream leading to a brackish lagoon surrounded by land. Dozens of *Caulerpa* clumps were rooted in the channel and at the entrance to the lagoon. Our ignorance of the source of the introduction (where we should have found a denser concentration of *Caulerpa*), the great dispersion of the clumps, and the presence of strong currents (which aid dissemination) led us to a pessimistic prognosis: the invasion of the small patches of bottom at Messina was going to proceed rapidly in the next few years.

Shortly after the announcement of this distant beachhead, an unusual person joined the assessment, and his activities became the talk of local columnists. Giuseppe Giaccone, phycologist at the University of Catania (Sicily), was also known for being a defrocked priest and the former mayor of a municipality in the suburbs of Palermo. He was, at the beginning of the 1990s, confronted during his term as an elected official with incredible Mafia business associated with the granting of public market licenses. The assassination of two of his friends forced

him to turn himself over to the legal authorities and to denounce the **169**
activities of a local Mafia clan. Suspected of wrongdoing before being
cleared, he was exiled for two years for security reasons to Rome, where
he was guarded day and night by armed men. This entire affair had
never stopped him from writing many normal scientific articles on the
marine flora of his country. As soon as the Sicilian newspapers reported
the arrival of *Caulerpa taxifolia* near Messina, he arrived on the scene
heralding the theme—already trite in France and Monaco—"no alga
ever killed anybody." [100] To counteract the terms "killer" and "assassina,"
he found good qualities in the beautiful tropical alga, thereby contradicting his phycological colleagues from Palermo (who had in any event
publicly denounced the use of excessive adjectives to describe the alga).
His eagerness to defend the alga in the media was certainly noticed by
the defenders of *Caulerpa* at the Oceanographic Museum of Monaco.
He was invited to collaborate with the Oceanographic Observatory of
the Scientific Center of Monaco (headed by Jaubert) in undertaking
studies financed by the Principality of Monaco. He too adopted the hypothesis that the alga came from the Red Sea via the Suez Canal. Ever
since, a heated debate has divided Italian oceanographers on the subject
of *Caulerpa* in colloquia of the Italian Society of Marine Biology, the
national press, and Italian dive magazines.

The only Italian community to participate in financing research on
Caulerpa taxifolia in Italy has been the province of Imperia. The division of responsibility for the marine realm is especially complex in Italy:
several ministries and agencies are in open competition over who has
administrative jurisdiction for maritime affairs. The overly split responsibilities and the disagreements among scientists have led to inaction;
there is no global strategy at the national level for fighting or studying
Caulerpa taxifolia. As in France, they let the alga grow.

At the end of 1994, the Italian experts estimated the area of bottom
patches invaded by *Caulerpa* in their country at 200 hectares, the majority (90 percent) found on both sides of Imperia. In 1995, the alga spread
to the west of Liguria, in a radius of two kilometers around the harbor
of Imperia, and it proliferated in the small neighboring harbors. In

170 1996, we estimated the area invaded by the alga in the region of Imperia at more than 1,300 hectares. Off the islands of Elba and Sicily, the alga has also spread from the two sites noticed in 1993. Finally, it was found again near Leghorn.

A novelist inspired by the affair had a certain success with the publication of his novel, *The War of the Basil.*[101] His story relates the inquiry by an Italian student into the algal invasion. She describes all the difficulties and pitfalls encountered in her research on the origin of the alga at the Oceanographic Museum of Monaco. It is easy to recognize in the main characters the different players in the real history. To spice up his story, the author introduces a torrid romance between the student and a fetishist searching for objects that belonged to the late Princess of Monaco, Grace Kelly. In the hotel where the two main characters describe their respective quests, a basil plant grows in monstrous fashion ... It's best just to laugh at it!

THE SPANIARDS FIGHT IT

The attitude of the Spaniards towards *Caulerpa taxifolia* impressed me greatly. As soon as the threat was publicized in France, the Spanish national authorities organized a meeting and invited the IFREMER experts and Boudouresque. The assistant director-general of the Spanish merchant marine was invited to France to attend the meetings of the Coordinating Committee on *Caulerpa taxifolia* created in February 1992 by the Secretariat of State for the Sea. He came himself rather than send a representative and related the intention of the Madrid authorities to create a Spanish national coordinating structure and a regional structure in Catalonia.

In August 1992 a private underwater works company (Drago-Sub) based in Majorca was assigned, by the Balearic region, to film the situation on the Côte d'Azur. Divers from my laboratory escorted the Spanish divers to the site where the invasion was then most spectacular, Cap Martin. Arriving from the Balearic islands on board a magnificent yacht transformed into a diving base, the Spaniards had sophisticated underwater cameras and video robots. A local television station broadcast their report.

Twenty days later, a Spanish diver of French ancestry, Nathalie Laffite, found 200 square meters of *Caulerpa taxifolia* in the midst of an anchorage heavily used by yachtsmen at Cala d'Or, on the east coast of the island of Majorca. Emotions of local fishermen were so inflamed that the private company that had contributed so much to alerting divers was assigned to eradicate the *Caulerpa* with the assistance of regional scientific advisors. The bureaucrats of the office of fisheries for the Balearics were competent scientists and experienced divers. Several days after the discovery of the alga, at the beginning of September 1992, the inlet was closed to navigation. An accurate map was drawn up of the clumps of *Caulerpa*, which were already scattered over a hectare. The divers next demarcated the bottom with string. The task of manual eradication began in the month of October. Two divers went over every demarcated square with a fine-toothed comb. When a clump was found, one diver tore it out, taking great care to remove every column of roots so that nothing was left. Another diver monitored the operation with a net, collecting the tiniest escaping fragments.

Invited to attend the end of this operation, I dove with the eradication teams. The plan of action was the same one that could have been used on the Côte d'Azur at the beginning of the invasion. I congratulated the local authorities for their rapid and efficient intervention. Two hundred square meters, dispersed over a hectare, were cleaned in two weeks. The inlet nevertheless remained closed to navigation and a watch was organized to track resprouting. The following summer, divers devoted a hundred hours to monitoring the site. Many resprouts, in the form of small, isolated plants, were again eradicated. In 1994, the divers found only a dozen resprouts. An assiduous examination of the site was planned until the total disappearance of the alga was verified. The method was shown to be effective: it had cost the Balearic authorities nearly 500,000 francs.

Unfortunately, in 1995, when the situation was believed to be under control at Majorca, more than 1,000 square meters of scattered clumps of alga were discovered spread over several hectares in an inlet and harbor in the vicinity of Cala d'Or. Everything had to start over again. Determined, the Spaniards decided to act once again. Divers of the civil guard were requisitioned to map and to eradicate the alga. Four hundred twenty kilograms were painstakingly ripped out in three months of diving. At the end of 1996, many colonies were still found over an area estimated at 20,000 square meters. Control of the algal spread in the neighborhood of Cala d'Or became more and more difficult. To prevent new colonization in the Balearic archipelago, maintenance of all exotic species of *Caulerpa* in aquaria was prohibited. But the most likely culprit for its arrival in Majorca—pleasure craft—remains uncontrollable. Thus, the marvelous bottom patches of these islands, swarming with fishes, remain very exposed to invasion.

On the eve of the summer of 1993, under the auspices of the European publicity campaign, brochures in Castilian Spanish and Catalan were distributed in all harbors and dive clubs on the entire Spanish coast. The Ministry of Education and Culture, the Ministry of Public Works and Transportation, and the autonomous governments of the Balearics, Valencia, Andalusia, Catalonia, and Murcia co-financed the publicity campaign, conducted under the aegis of the Spanish Institute of Oceanography and of the Higher Council of Scientific Investigation. To prevent any new introduction of the alga by aquarists, Catalonia prohibited, six months before France, the transport, sale, or possession of the alga.[102] Scientific divers inspected Catalan harbors and anchorages. Despite several false alerts, no *Caulerpa taxifolia* was found on the mainland coasts of Spain.

Two teams of Catalan researchers (from the University of Barcelona and the Oceanographic Station of Blanes, north of Barcelona, respectively) participated in the European research program on *Caulerpa taxifolia*. One worked in particular on many biological traits of the alga (its oxygen production as a function of light and temperature, its requirements for nitrates and phosphates); the other conducted research on the effect of *Caulerpa* toxins on certain Mediterranean algae.

It was in this context that the reappearance of the alga in Septem-   **173**
ber 1993 at Saint-Cyprien, some thirty kilometers from Spain, irritated
the Spanish authorities. In March 1994 the Spanish, worried by the
spread of the alga so close to its coasts, asked the French authorities to
suppress it. In the spring of 1994, IFREMER drew up a cost estimate
for the eradication of 800 square meters of *Caulerpa.* The summer of
1994 passed; the alga again grew without being inconvenienced in any
way. At the end of November 1994, we had to map the spread of the alga
at Saint-Cyprien. A team of seven divers from our laboratory was able
to establish that 1,250 square meters were entirely covered by the alga
and scattered cuttings spread out over another 1,000 square meters. The
eradication bill had to be raised. The consequences of the hesitation of
the decision-makers can be calculated exactly. And while the government experts calculated the new cost of eradication and sought new
funds to rip up the alga, it continued to grow. By the end of 1996, more
than 4,000 square meters were invaded in the harbor.

THE CROATS APPEAL TO THE EUROPEAN UNION IN VAIN

In February 1995, a researcher at Split brought to France a fragment of
Caulerpa taxifolia found off a Croatian Adriatic island. He gave it to
Henocque, of IFREMER, who in turn sent the sample to me. There was
no doubt; it was indeed *Caulerpa taxifolia.* The alga had just arrived in
the Adriatic Sea.

I quickly mounted an expedition with two collaborators to understand the location and appearance of the colonization. In March 1995,
after a long trip amidst humanitarian and military convoys, we were
greeted in the early morning by Croatian scientists. They immediately
took us aboard their oceanographic boat and we navigated for several
dozen kilometers towards one of the many islands that dot the Croatian
coastline. The alga had been found in front of a landing quay on the
island of Hvar. It had already invaded 6,000 square meters of small
patches of bottom. Although it was winter, the water was clearly

174   warmer than on the Côte d'Azur at the same time: 14.5 °C. The *Caulerpa*
had not suffered much from the cold and seemed already to be spreading in its conquest of available space. The alga began to penetrate the
Posidonia meadows and covered the *Cymodocea* (another marine flowering plant genus). At Hvar as at Split, the local scientists organized
meetings, with our participation, to inspire divers to search for other
possible colonies.

A second extension of *Caulerpa* had been reported 200 kilometers
further north in the Adriatic, on the coast of the island of Krk, located
not far from Rijeka. The roads connecting Split and Rijeka bordered
Krajina, the Croatian province then controlled by the Serbo-Croatians.
We skirted that dangerous area by ferry and joined another team of
Croatian oceanographers that had identified the alga in that region. It
covered 600 square meters of muddy bottom in the little harbor of Malinska. The water there was very cold, 10°C, and the site is north of the
45th parallel; it became the northernmost site in the world where an
alga of the genus *Caulerpa* can be found.

We tried to understand the presence in the Adriatic of the cold-
resistant strain of *Caulerpa*. The most plausible hypothesis would be the
accidental transport of cuttings on a boat anchor. A small tramp steamer
coming from Messina or Liguria could have dropped cuttings of *Caulerpa* during calls at these two sites two or three years previously, before
the Yugoslavian conflict. In the north of the Adriatic the water is cold;
the alga could grow there only three or four months each year. Further
south, towards Split, the alga could spread for five or six months every
year, which could explain the difference in colonized area between the
two sites.

The Croatians counted heavily on our aid. Delegates from the Ministry of the Environment participated in the discussions. We had given
them all the available scientific documentation on the subject and had
discussed at length the best strategies to try to fight or to control the
situation. But the local scientists were especially powerless in the face
of this problem. The Croatian authorities appealed to the European

Union for assistance to fight the alga. To my great regret, their request
was not granted. While an attempted eradication of the alga in the harbor of Malinska failed at the beginning of 1996, a third colony was discovered in an anchorage (in the Rab canal).

RETURN TO MONACO

At Monaco, everything is under the authority of His Royal Highness
Prince Rainier. This situation does not prevent heated battles between
opposing factions on anything related to power associated with government management. The problem posed by *Caulerpa taxifolia* proved to
be no exception. Matters of the sea are under the supervision of three
organizations with overlapping jurisdictions: the management of the
property of the marine reserve of Monaco is granted to the Monacan
Association for the Protection of Nature (AMPN); the management of
the marine environment is given to a state agency, under the supervision
of the Ministry of Equipment; the scientific surveillance of the marine
habitat is assigned to the Monacan Scientific Center, and a large part
of the research facilities of the latter are housed in the Oceanographic
Museum of Monaco.

The president of AMPN, a former bureaucrat of the Ministry of the
Interior of the principality, has never disguised his total lack of expertise
in matters relating to marine biology. Not a diver himself, he nevertheless presided with spirit, devotion, and passion over the association since
its creation. He thus did a good job in the main mission confided to
him by the Prince, which was to show the marine reserve of Monaco to
advantage. This is a protected area that extends over 50 hectares in front
of the artificial, highly urbanized coast of Monaco. The drastic prohibition of fishing and the establishment of artificial reefs have attracted
the fishes that remain in the reserve. They no longer fear humans, and
it is very easy to approach them. To snorkel with a facemask along the
artificial beaches of the reserve is a true delight. The result of his twenty
years of management is remarkable. The only shadow on this picture is

176   *Caulerpa*, which greatly threatens the wondrous biodiversity of the reserve.

I have often dived between the Museum and the reserve. Since
1990, each summer, I have swum over several kilometers, scrutinizing
the bottom through my mask. Holding my breath, I dove down to many
suspect green spots glimpsed from the surface. I have measured the
Caulerpa colonies and noted their locations. I have thus observed the
slow eastward progression of the alga, against the current. In 1991,
the alga was attached to the rocks of a dike 300 meters from the reserve;
in 1993, it was in the reserve. At the end of 1995, several dozen clumps
from three to six square meters were strewn over the *Posidonia* meadows of the reserve. In my explorations of this region, I have photographed everything and transmitted everything to the president of the
association. Beginning in 1989, I had warned him about the danger. He
often discussed with me his efforts to communicate this subject. All the
information I sent him reached high places through his efforts. As soon
as the alga arrived in the reserve, he had my observations verified by
the Monacan firemen, who also explored the whole bottom of the reserve. Their observations confirmed the advance of the alga. Convinced
of the danger, the president of the association never succeeded in persuading the higher authorities. In November 1995, for the twentieth an-
niversary of the creation of the reserve, he hoped to bring together the
most beautiful underwater pictures in a synthesis of the most notable
research conducted in the protected zone. The album is magnificent and
its production luxurious. But it includes an article by Jaubert that, while
presenting the activities of his Oceanographic Observatory, includes two
misleading pages about *Caulerpa taxifolia*. The president of the association was forced to accept this article, just as he was forced to accept
Caulerpa in his reserve. How did he feel when he saw the arrival of the
alga that is proceeding gently and inexorably to cover the fruit of twenty
years of labor?

I knew the chief of the environmental agency of Monaco well. We
had pursued our studies together at the University of Nice. He special-

ized in teaching the natural sciences in a high school in Monaco, then   **177**
joined the administration in order finally to indulge once again his true
passion, the sea. He was the one who represented Monaco at some meetings of the Coordinating Committee on *Caulerpa taxifolia.* He undertook to monitor the invasion on all Monacan coasts. In 1992, his agency
funded a mapping of the entire sea bottom of the principality, by means
of a robot furnished with a camera and connected by an umbilicus to a
ship on the surface. The patches of *Caulerpa* observed by chance during
his trajectory were mapped. The Monacans were the first to know that
the alga lived at a depth of seventy meters. The chief of the environmental agency made the Prince's police officers dive and regularly inform
him of the progress of the alga. The agency chief subsequently collaborated with IFREMER on a detailed cartography of the alga in the principality. I imagine these reports were also transmitted to the highest
authorities. But in taking the initiative and implanting *Posidonia* in the
underwater reserve, the chief of the environmental agency revived a
conflict between his agency and the Monacan Association for the Protection of Nature. Although they had similar positions on the problem
posed by *Caulerpa taxifolia,* the two organizations were too far apart to
take joint action on this matter.

The scientific monitoring of the bacteriological quality of Monacan
waters has always been conducted by the Monacan Scientific Center.
Recall that the European Oceanographic Observatory, arm of the Monacan Scientific Center and housed in the Oceanographic Museum of Monaco, is under Jaubert's direction. We will return to the efforts of the
researchers based in the Museum to defend their opinions on the origin
and impact of the alga. At a high level, the synthesis of diverse opinions
has always leaned toward the side of the defenders of *Caulerpa.* On
many occasions, I have written to the Prince and tried in vain to meet
him to explain the situation in person. He has always responded courte-
ously to me but has never received me. He has sent me to his subordinates who have listened politely. One of his responses shows his position
clearly. He informed me that the director of the Oceanographic Mu-

178   seum of Monaco had kept him "informed of the presence of a verdant
       prairie of *Caulerpa* that has colonized, at the foot of his establishment,
       the former putrid expanse of mud formed by deposits arriving with the
       effluent of waste from the slaughterhouse." [103]

Research Progresses...and the Polemic Persists

Badly in need of funding, research marked time in 1992. The only studies consisted of the mapping of areas touched by the alga. The maps made a growing number of the people we spoke to admit that *Caulerpa* was spreading continuously—the key to the whole problem. We had to wait until 1993, when, thanks to the support of the European Union, nearly 150 researchers in thirty laboratories were finally able to study all aspects of the problem. A colloquium on *Caulerpa taxifolia* at Barcelona the following year provided the occasion to draw up an overall balance sheet. It ended with a unanimous statement that the invasion of the alga was a major threat to the coastal ecosystems of the Mediterranean. But this warning was not heeded by the institutional and political authorities; from 1995 to 1996, the problem was somewhat neglected. During this time, the alga continued to grow and to spread and the polemic with the defenders of *Caulerpa taxifolia* recrudesced in all its glory.

The Scientists at Work

OBSERVATIONS OF A DIVER

It is early July 1992. I dive over a *Caulerpa* bed. The water is 20°C at the surface. At this time of year, the Mediterranean algae barely begin to grow between the surface and five meters deep; in this zone, highly exposed to the movement of the water and temperature variation, *Caulerpa taxifolia* has regressed in the winter and the spring. Below, the *Caulerpa* prairie is more stable, but the cold and the winter storms have modified its appearance; the majority of fronds are broken or shriveled and have multiple branch points. The patches of bottom most exposed to the undertow and the peaks of large blocks of rock are stripped of their *Caulerpa*. But the algae closest to these spots already direct their creeping stolons towards these areas of recolonization, engaged in a battle to reconquer their territory.

Deeper, the *Caulerpa* vegetation is more regular and spreads out over the entire rocky surface. From this prairie emerge thousands of stolons bearing several small fronds. These stolons, new this year, go a little further, in the areas populated by old *Caulerpa* that have overwintered. The new vegetation prepares to replace the old. In two months, all the old fragments will disappear. The alga, consisting of a single giant cell, constitutes a new "body" identical to its predecessor. It is as if the living substance (the liquid interior, or cytoplasm) passes from the old fronds to the new ones. Once this "moulting" is finished, the old parts, emptied of their "green blood," become blanched and quickly dissolve. Thus the individual appears eternal, even though its physical components (the stolons, rhizoids, and fronds) are ephemeral; it renews itself each year. This strange cycle of vegetative growth is found only in the Caulerpales, an order comprising nearly two hundred essentially tropical species. During the "moult," the prairie always remains dense, green, monotonous, undulating by successive waves under the influence of the swells. One frond replaces another. When it is seen from afar, it is impossible to comprehend what is happening.

On the clumps of *Caulerpa*, ocellated wrasses construct their nests, especially between three and six meters deep. The male delicately takes algae and deposits them to construct a mass in the form of a bowl. With his multicolored finery, the beautiful male then parades over his piece of work. It is the nuptial period, and he awaits females. Small, more drab, the many candidates swim around him. Copulation is fleeting: side by side, facing the same direction, they brush their undersides together and simultaneously release eggs and sperm at the bottom of the nest. Everything appears normal, except the nest is not constructed of the invading species that henceforth dominates the landscape, but of the algae that these fish love, brown algae in the form of fine strips (*Dictyota* or *Dilophus*). They seem irreplaceable. Fortunately, they are still available around the *Caulerpa*.

A little further away, an octopus sits on the *Caulerpa*; I have seen it and it has seen me. It hides in the short *Caulerpa* prairie, but it does not succeed in taking on the fluorescent green color of this new décor, so it is not hidden from the danger I represent. On *Caulerpa*, mimicry is no longer possible. The octopus takes flight, protected by an ink cloud that it releases before escaping. From afar, I spy a red mass on the green *Caulerpa*—a scorpion fish. It too can no longer use its usual palette of colors to melt into the background and thus to lie in wait to surprise its favorite prey, small crustaceans or young fish. Are scorpion fish and octopuses on their way out in this site, where nothing resembles what they have always known? Or will they succeed in adapting to this new environment? How will they be able to capture their prey without being harassed themselves by other predators?

I descend towards the depths, where twilight interrupts the darkness for only a few hours. I know that there is a beautiful underwater cliff between twenty and thirty meters of depth, and vertical for more than ten meters. Multicolored in the beam of a flashlight, it was magnificent. It harbored an ensemble of species adapted to that peculiar environment. Calcareous red algae and, above all, red and yellow gorgonians were the most visible living elements of this unique ecosystem, where hundreds of species of algae, invertebrates, and fishes also lived.

182 Lobsters, moray and conger eels, fork-beards, red corals, and small, beautiful red fishes (cardinalfishes and basslets) also abounded. The richness of Mediterranean underwater cliffs is a main attraction for scuba divers. Imagine how sad I was to find my cliff now a monotone green! Cascades of *Caulerpa* covered everything. Only several tips of the most beautiful gorgonian branches emerged from the green carpet. Gorgonians are animals that live attached to the rock; they are colonies of small polyps secreting axes or rods, analogous to the branches of plants, by which they spread. These arborescent axes, several dozen centimeters long, tremble in the current, thus allowing the minuscule polyps to trap organic particles or tiny living prey that flow past the cliff. Under the *Caulerpa*, there is no more current. The gorgonian axes covered by the algae have lost their polyps; only those located on the tips of the branches surmounting the *Caulerpa* are still functional. The gorgonian larvae, descending from these "marine flowers," will no longer choose this hostile environment for their development. The ecosystem has changed.

Among the other species affected by the altered ecosystem, the edible sea urchin, exclusively vegetarian, hardly even samples this venomous vegetation, especially in the summer. On the other hand, the damselfishes and blotched picarels, those little fish that always swim in scattered reefs and hunt in the open water, appear insensitive to the difficulties of those who need the fauna and flora of the bottom patches to survive. And the others—the multitude of anonymous organisms, for which much less (or nothing at all) is known about their food habitats, reproductive circumstances, and normal natural histories? How will they be able to adapt to the tropical alga that is so lush and green, but inedible? For each animal species, for each Mediterranean alga or plant, I ask the same questions. How will they make a living in this new, wholly green universe? Which species will be eliminated, which will be threatened, and which will adapt? How will the many and varied ecosystems found on the Mediterranean coasts resist the change? What will be the overall change in the biodiversity?

At times these seem futile questions, useless for those who live

above, outside the water. Have I become an environmental fundamen-
talist, lover of underwater cliffs or beautiful gorgonian corals? A partisan
of "deep ecology," the view of some ecologists that mountains, trees, and
animals have spirits? No, I have not. Whether or not radical environmentalists or advocates of animal rights like it, I have enjoyed hunting
underwater for squid, octopuses, sea bream, scorpion fish, mullet, and
other wild game. I have not been very reckless in catching my prey.
I have never gone so far as to flush large fish out of deep holes, or to
remain motionless underwater, holding my breath, hidden between two
rocks to wait for the curious fish to come on its own. But the hundreds
of hours spent watching have taught me to observe the behavior of
different species; I have accumulated many insights that have allowed
me to understand each situation, each community, each species under
study. But I have retained the instinct of the hunt whenever I dive. Even
if I have replaced the crossbow with the underwater slate, the camera,
or the sampling bag, I often enjoy rousting out fishes in order to approach them very closely.

But the view of this great upheaval takes away my desire to hunt. I
am sickened. What is happening? What have we done? What can I do?
To whom should I communicate my distress? Each time that I dive at
this spot, I feel responsible. As soon as I saw the alga covering the
patches of bottom in front of the Museum, I understood what was happening and what was going to happen everywhere. It is maddening to
observe, month after month, that I was right, maddening to have encountered so many roadblocks while trying to make people conscious of
the drama playing out underwater.

SCIENTIFIC PROOFS

The invaded ecosystems are disorganized, they are disturbed, they
change. Any diver can make this observation. But it is up to the scientist
to prove it and to describe the precise impact of the invasion. Very early
in the affair, our team took on this challenge. All the publications on
the theme of *Caulerpa taxifolia* appearing in 1991 and 1992 bore the

184   name of at least one member of my laboratory. Several years passed before other research teams from various universities or institutes (CNRS, IFREMER, or INSERM) contributed to the effort. In total, nearly 150 researchers from 34 laboratories and institutes of three countries (Italy, Spain, and France) participated in the studies conducted between 1993 and 1995, financed essentially by the directorate of the environmental office of the European Union and effectively coordinated by Boudouresque.

An exhaustive synthesis of different advances in knowledge about *Caulerpa taxifolia* is beyond the scope of this work.[1] It will suffice here to summarize some of the proof accumulated that the invasion affects biodiversity and disrupts entire ecosystems.

Fish specialists (ichthyologists) tallied the fishes in different seasons along marked routes. They selected two routes invaded by *Caulerpa* and two routes without *Caulerpa*, or control routes. After monitoring these areas for two years, the researchers observed a significant drop in the sizes and numbers of individuals of certain species along the invasion routes.

Specialists in algae, molluscs, crustaceans, and worms scraped rocks invaded by *Caulerpa* until they were bare and inventoried what they had removed. In comparison with what lives on uninvaded rocks, they noted that the crash in biodiversity is especially marked among Mediterranean algae; it is more subtle among the small invertebrates. There is even an increase in numbers of individuals of certain worm and mollusc species that thrive on muddy bottoms. The *Caulerpa* prairie often constitutes a muddy humus that favors these animals.

Several studies were initiated on the consequences of the invasion of the *Posidonia* meadows that form underwater "forests" on Mediterranean bottom patches. A team of researchers from Marseilles detected a chemical battle between *Caulerpa* (which invades the meadows and exudes toxins) and the *Posidonia* (which produce tannins, a type of defensive compound, as soon as the *Caulerpa* approach them). The *Posidonia* seem to resist the invasion for a while. Their annual cycle is opposite to that of *Caulerpa:* their leaves grow from February to June, and it is during this period that the *Caulerpa* stolons carry only scattered, stunted fronds. By contrast, from July to January, the *Posidonia* lose some of their leaves, while the *Caulerpa* grow rapidly, invading the *Posidonia* meadows and covering them. Even the initial research showed that sparse *Posidonia* meadows were threatened. But in 1994 and 1995, several observations confirmed that even the densest meadows were also invaded. The changes wrought in this ecosystem have been substantiated by other research. When the *Caulerpa* density rises between the bundles of *Posidonia* leaves, hundreds of meters of stolons and tens of thousands of rhizoids per square meter cover the substrate. The *Caulerpa* fronds trap suspended mineral or organic particles, which circulate freely between the *Posidonia* leaves. Particles settle on the bottom, and this sediment is stabilized by the rhizoids. The presence of *Caulerpa* in the midst of a *Posidonia* meadow thus leads to silting, which causes profound changes in the fauna and flora associated with *Posidonia*.

The first contact between *Posidonia* and *Caulerpa* was observed at Cap Martin in 1990. It is therefore still much too early to predict the final outcome of this confrontation. Will the fields of *Caulerpa* replace the *Posidonia* meadows? Will islands of *Posidonia* resist the repeated assaults of *Caulerpa*? How well will *Posidonia* be able to regenerate and reproduce? Will *Posidonia* cuttings or seeds carried by currents be able to establish in the midst of *Caulerpa* prairies? Much other research still needs to be done.

Several teams have studied how *Caulerpa* toxins affect other algae, sea urchins, marine bacteria, and a dozen marine microorganisms. Their work shows the varied impact of the toxins on the environment. It currently appears that the toxins act as a repellant against the herbivore fauna and prevent the attachment of other algae and microorganisms to *Caulerpa*. But several marine species are quite resistant and grow normally. Three teams are investigating the chemical structure of the *Caulerpa* toxins. The toxicological research has shown that the main toxin (caulerpenyne) does not accumulate in food chains. Seafood grown on *Caulerpa* fields does not contain these toxins, eliminating any possibility of health risks from eating harvested species. But caulerpenyne is very

186   unstable; its breakdown products, also toxic, are still being studied.[2] The increased area occupied by *Caulerpa* and its high density perturb the ecosystem and force us to be on guard for toxic effects.

To attempt to understand how *Caulerpa taxifolia* comes to dominate all other algae, some researchers have begun physiological studies. What are the phosphate and nitrate requirements of the alga? Where does it get its carbon, the major constituent of all life? What becomes of the organic matter when the alga decomposes?

The amount of data needed to assess environmental change thoroughly is daunting. Scientists often refer to their conclusions as weak, because they are based on inadequate data, and the time factor is also important—the many components of the invaded ecosystems have varying degrees of resistance. Moreover, the study of all the changes wrought by the invasion is hampered by the small number of scientists available to study the effects. In the future it will be difficult to find European specialists competent and willing to embark on the study of the adaptations of many groups of species.

To anticipate changes involving interactions among thousands of species is a risky business. Because of various reproductive traits, life-spans, or eating habits, some species will persist longer than others. Some will adapt more or less quickly to the new situation. But how many will disappear? It will take years to assess the potential affects of this invasion. It is as if a mimosa tree, of matchless vigor and resistance, grew on the seashore and was advancing towards our highest mountains. What would remain of our Mediterranean scrubland, our moors, our oak groves, our fir forests, and our alpine meadows? How well would these ecosystems and their component species resist the invasion? With *Caulerpa taxifolia*, these events are happening underwater, which makes observations and research much more difficult.

In research on a disturbance of this sort, one must focus on the themes that will answer essential questions and permit us to foresee the final impacts of the invasion. Two subjects in particular require in-depth analysis. The first is the progress of the alga and the prediction of how far it will spread. Knowledge of the dynamic of the invasion is the key

to the problem; if the alga ceased to spread, all of the costly investments
in research on the ecosystem would be unjustified. Laboratory research
on the factors limiting this spread has failed to instill optimism—neither temperature nor light nor dissolved nitrates or phosphates typically
found in seawater (the principal nutrients for plants) will limit the progression of this alga throughout the Mediterranean. We know that a
single cutting can lead to the invasion of ten hectares in six to seven
years. Computer scientists at the University of Clermont–Ferrand have
produced a program that allows one to visualize the invasion of a bay
beginning with a single cutting and predicting the trajectory of the
invasion for five years. Simulations on a larger scale are under development.

The second crucial question concerns the genetics of the species.
The accumulated data tend to show that the alga invading the Mediterranean differs from the *Caulerpa taxifolia* of tropical seas. After five
years of observation, no sexual reproduction has been noted in the Mediterranean stock; when an individual alga transforms all its nuclei into
gametes, only sperm are found. Instead of being hermaphroditic, as in
tropical waters, *Caulerpa taxifolia* individuals in the Mediterranean are
all male! Until there is proof to the contrary, one can assume that all
the Mediterranean specimens are thus offspring of a single individual
fragmented into millions of genetically identical cuttings. The reproductive traits, the vigor, and the cold-resistance of this clone are very
different from those of its tropical "cousins." These differences can be
explained by a change in the genotype of the alga. Was there a mutation
in one or several chromosomes? Was there a doubling of chromosomal
number?

To try to answer these questions, we have been growing in aquaria,
since 1992, samples of *Caulerpa taxifolia* collected in various tropical
locations and in the Mediterranean far from Monaco. A lack of sufficient research funds prohibited detailed analysis of this collection until
1997.

If one of these hypotheses were to be confirmed, we could be confronting the first marine invasion of an organism genetically different

188 from those encountered to date in nature. It was certainly not created intentionally by humans, but somehow this species changed in aquaria. The alga was selected by aquarists for its extraordinary qualities of survival, vigor, and absence of sexual reproduction (sexual reproduction of tropical *Caulerpa* species, which occurs frequently in aquaria, leads to their death). Selected by humans, this clone was, accidentally or not, released into a natural habitat, the Mediterranean. Its aggressive proliferation was not stemmed in time, and it is now spread by fishing boats and leisure craft. The role of humans has been crucial at each stage and remains so. This alga could invade elsewhere—the coasts of the entire Mediterranean Sea, but also the tropical, subtropical, and temperate coasts of oceans communicating with the Mediterranean, including the Red Sea and the Atlantic Ocean.

Other hypotheses have been advanced to explain the spread of *Caulerpa taxifolia* in the Mediterranean. One theory links the invasion to global warming, which would favor the spread towards the northern Mediterranean of species until then confined to the warmer, southern shores of the eastern basin. This hypothesis is not credible. It has in fact been demonstrated that the clone that is proliferating in the Mediterranean survives three months at 10°C; it is thus not a tropical species but one adapted to temperate climates. It would have grown just as well if the temperature of the Mediterranean had dropped. It is impossible therefore to connect the invasion to global warming.

The substantial increase in the number of species introduced to the Mediterranean has also been frequently cited as another theory. The majority of these recently introduced species have not thrived in their new habitat. Many introduced species disappear quickly. Some persist by finding an unoccupied niche without spreading beyond it or by replacing an ecologically equivalent species. Those that proliferate are rare, and those that have a tendency to expand are usually restricted to one ecosystem. But above all, none have the vigor and potential of *Caulerpa taxifolia*, a species that is dominant, persistent, and ubiquitous, able to adapt to a great diversity of ecosystems.

Pointing out the many recent introductions tends to minimize the problem posed by the invasion of the most damaging species. By the precautionary principle, we should attempt generally to limit introductions. It is doubtless this recommendation, which we made in 1991, that was the most influential. Since 1995, a French law stringently regulates the introduction of species.[3] During a meeting of ambassadors on the convention to protect the Mediterranean from pollution, a protocol was established including two recommendations on species introductions that are clear and allow no exceptions.[4] The first states that all appropriate measures will be taken to regulate the deliberate or accidental introduction into nature of nonindigenous or genetically modified species and to prohibit those that could have harmful impacts on ecosystems, habitats, or species in the zone of application of the protocol. The second mandates the use of all possible measures to eradicate species that have already been introduced when, after scientific evaluation, these appear to cause or to be capable of causing damage to ecosystems, habitats, or species in the zone of application of the protocol.

In various international settings, Boudouresque figured out how to generate interest in the worrisome and unique *Caulerpa taxifolia* situation. At the level of the European Council, in the context of the convention on conservation of wildlife and natural habitats, a recommendation was adopted calling for the control of the spread of *Caulerpa taxifolia* in the Mediterranean.[5] The nations bordering the Mediterranean were called upon to collaborate to map and eradicate the alga locally in order to halt its spread. The same conclusions were drawn under the aegis of the United Nations in the context of the action plan for the Mediterranean.[6]

At the Barcelona colloquium in December 1994, the researchers engaged in the European program launched an appeal to relevant authorities, entitled "*Caulerpa taxifolia:* confirmation of a major risk to the coastal ecosystems of the Mediterranean," which summed up the results of two years of studies and sounded the alarm bell: "Even if we cannot yet predict all the consequences of the spread of *Caulerpa taxifolia* for the coastal environment of the Mediterranean, and even if some hypotheses have not been verified, the data gathered to date confirm

190   that there is a major risk for biodiversity, ecological equilibrium, and exploited resources.

"The scientists are doing their job of research and continue to take responsibility for alerting the authorities. It is now up to the governments of the concerned countries as well as international organizations in charge of the environment (UNEP, the Barcelona Convention, IUCN . . .) to establish a precautionary principle (as called for by the Rio de Janeiro Convention) and to define a coherent international strategy suited to the problem at hand."

The protocols, laws, recommendations, and other appeals did not greatly change the situation. The scientists turned in their papers after they exhausted their research funds and waited for the politicians to take over. Nothing happened.

Even without any contractual support, our laboratory continued to map the invasion. For every new discovery, a report was sent to relevant governmental authorities. Beginning in 1995, we systematically added a warning to the observations: "if you do not eradicate the several square meters that have been recorded, the situation will quickly go out of control in the sector that has just been invaded." None of the authorities to whom we sent this information acted on it, and we never even received an acknowledgment that our report had been received.

In 1995, Boudouresque sought to involve several research teams in the establishment of two European programs of study of *Caulerpa taxifolia*. The first focused on mapping, eradication, and reports for countries outside the Mediterranean basin, and it was accepted in September 1996. The second was a question of fundamental research, aimed particularly at better evaluation of the impact on biodiversity and at conducting a genetic analysis of the invading *Caulerpa*.

The media sowed seeds of doubt among the European Union experts, and the program was rejected. It was our fourth attempt to fund a genetic study of *Caulerpa* and our fourth failure. Inertia or official contempt, as well as the efforts of our adversaries to support their hypotheses, would have discouraged even the most stubborn individual.

But a dive over the *Caulerpa* prairies sufficed to inspire us to persist in 191
our attempts to convince others of the impact of the invasion.

The Rearguard Battle of the Scientists Based at the Oceanographic Museum at Monaco

While several teams of scientists worked on different research themes, the directorates of the Oceanographic Museum and the European Oceanographic Observatory of Monaco strained mightily to keep up the polemic. From 1992 to 1996, every available means was used to feed the beast of controversy: media, legal, and scientific.

"CAULERPA—HAVEN OF LIFE"

At the end of July 1992, the strategists based at the Oceanographic Museum mounted a media campaign with the support of the International Commission for the Scientific Exploration of the Mediterranean (CIESM). Reporters were invited to the Oceanographic Museum to learn of the activities of Prince Rainier at the United Nations conference on the environment and development, which had just been held at Rio de Janeiro. Obviously, it was really about *Caulerpa.* A ten-minute film was shown, entitled "*Caulerpa*—Haven of life," on the theme that "the *Caulerpa* prairies are oases of life." Produced by the Museum, the film was a montage of several sequences showing fishes and octopuses swimming amidst *Caulerpa.*

Under the gilt paneling of the prestigious Prince Albert I Lecture Room, Doumenge and Jaubert discussed the film. They took turns refuting en masse all the fears about *Caulerpa taxifolia* and emphasizing the environmental benefits of the alga. A third participant, Frédéric Briand, who a few months previously had become director-general of CIESM, gave the backing of that international institution.[7] The film proved nothing, for one could as easily have produced a different one, of ex-

panses of *Caulerpa* deserted by fishes. But it was a high-quality production, and its promoters counted on their high-ranking positions to
make their case. The support of the CIESM, the influence of which was
strongly emphasized (it is a federation of some 2,500 researchers from
Mediterranean countries), must have impressed people. The scenario,
even if scientifically barren, bore fruit. Once again the "killer alga" surfaced in the press.

"Eminent scientists see fabulous ecological virtues in it," read a sub-
headline in *Le Provençal*.[8] The alarmist researchers were taught a les-
son. "A scientist should always seek the truth; his goal is not normally
to erect a scarecrow under the spotlights," declared Briand, who even
suggested the utility of deliberately introducing *Caulerpa* in certain locations where no marine flora had managed to grow. According to him,
"it seems, after months of research, that *Caulerpa* plays an interesting
and beneficial role. It reclaims degraded, polluted muddy bottoms."[9]

For his part, Jaubert announced imminent research results proving
that *Caulerpa* lives exclusively in polluted habitats. Although just several days earlier he had rejected the invasive character of the alga,[10] this
time he recognized its spread, associated in his eyes with the increase in
atmospheric carbon dioxide. However, Jaubert was no more able than
Doumenge and Briand to cite published research.

The chair of the "algal subsection" of CIESM had not been consulted. By the same token, none of the four hundred specialists in the marine flora and fauna of Mediterranean countries (members of CIESM)
had given their support on the subject of *Caulerpa* to the director-
general of the commission. His advocacy of an arbitrary and dogmatic
position was subsequently heavily criticized by many researchers. Because CIESM was financed by the governments of the member countries (all the countries of the Mediterranean) and France was the principal source of funds, I interceded, to no avail, before the supervising
French ministry (the Ministry of Foreign Affairs) to express my astonishment at such a partisan attitude. Despite these reactions, the video
"*Caulerpa*—Haven of life" was distributed to reporters and participants in the debate.

August 1992–June 1993: The Lawsuit

In the course of a widely viewed regular broadcast, the television network Antenne 2 showed a report on *Caulerpa taxifolia*. The show emphasized the brewing ecological drama and cited the two opposing hypotheses. The producers showed a map of the Mediterranean depicting the route the alga had taken, according to the Museum researchers. Leaving the Red Sea, the alga would have traversed the Suez Canal, then skirted the meanders of the tortured geography of the Mediterranean shores to arrive at Monaco after a journey of several thousand kilometers. In this exposition, the hypothesis defended by the Museum seemed ludicrous compared to that of "the accident"—the alga, cultured in the Museum aquaria, would have been purged into the Mediterranean. Apparently this report was not appreciated at the Museum, to the extent that the Oceanographic Institute sued Antenne 2, its director Hervé Bourges, and the producers of the broadcast film—Paul Nahon, Bernard Benyamin, and Richard Binet—for libel.[11] The invasion of the alga, described in the broadcast as a catastrophe, was, according to the Institute lawyers, ascribed too directly to the Oceanographic Museum of Monaco, which pleaded it had suffered "substantial damages."

The trial lasted six months. For this whole period, the defenders of *Caulerpa* tried hard to convince the journalists that the scientists involved in the affair had acted out of jealousy, rivalry, or self-interest in accusing the Museum. This was the most lamentable episode of the "cold war of the killer alga." Jaubert was personally heavily involved in these legal maneuvers. In the local press, he defended as his own the ideas propounded by Doumenge on the origin of the alga, and he insisted that *Caulerpa* was benefiting the environment. His photograph appeared in a local newspaper under the headline: "An advocate of *Taxifolia.*"[12] "Prof. Jaubert: it is science fiction" heralded another newspaper.[13] He announced the publication of definitive research: "*Caulerpa taxifolia* was not in our planned research program, but faced with such a media drumbeat and the alarmist statements of some of our colleagues, we developed a physiological study with the goal of determin-

194 ing the factors that govern its growth." He strove above all to denounce the alarmist scientists as troublemakers for having supported the apocalyptic scenario in the telecast, accusing them of "knowingly fostering the psychosis of the 'killer alga.'" [14]

This facile accusation was warmly echoed by the local authorities—finally a sober and reassuring scientist who denounced his colleagues' outbursts. But he did not stop at this deceitful exploitation of the media hype. He suggested to reporters that we could have exaggerated the whole affair. One of them received a remarkable letter from Jaubert inviting him to "look a bit beyond the tips of the *Caulerpa*. The firestorm provoked by the sulphurous '*Taxifolia*' case might betray a malaise among a certain class of biologists. It also might have aspects that go beyond ecology."

One reporter, Franck Jubelin, was especially captivated by such suggestions. The geographic proximity of the protagonists of the "*Caulerpa* affair," [15] an ambiguous and confidential document [16] sent to Jaubert by the director of the faculty of sciences, his confidential statements and those of two of his senior lecturers all convinced him that I had staged the entire affair in order to accuse the Museum of negligence, and that my motive was jealousy. Jubelin's analysis appeared in the monthly magazine *Thalassa*, presented as a scoop and headlined "The mysterious fourth floor." [17] I tried in vain to get my university administration to weigh in against this underhanded attack aimed at my professional reputation. By refusing to respond to my written requests to obtain an official summary and clarifications, my superiors did not allow me to discredit the version of the facts presented by the defenders of the Museum. The affair of the "mysterious fourth floor of the faculty of sciences" was recalled in his plea by the lawyer for the Museum. He cited the *Thalassa* article and evoked my supposed jealousy towards and rivalry with the Museum laboratories.

In February 1993, we were surprised to learn that Antenne 2 had been convicted. The conviction, of course, quickly became a big media story with the help of the Museum. [18] But Antenne 2, which became

France 2 in the interim, appealed and the conviction was reversed;[19] the television network was removed from the case and the producers were found not guilty of the crimes with which they had been charged. There was no further mention of this legal battle lost by the Museum, and the Museum had to bear the expense of the trial.

From then on, the most plausible hypothesis, the accidental introduction of *Caulerpa* by the Oceanographic Museum of Monaco, was presented as an obvious fact. Subsequently, radio stations could fearlessly broadcast a "rap" melody on the theme of marine pollution, with a verse saying: "*taxifolia*, you owe that to Monaco." In textbooks the invasion of *Caulerpa* into the Mediterranean from the Museum was the object of exercises in plant biology.

For want of support from the directors of the University of Nice–Sophia Antipolis in the affair of the "mysterious fourth floor," I had to defend myself alone in asking *Thalassa* for the right to respond, and my response appeared late (after the initial judgment in the lawsuit).[20] Hoping that the university administration would deal with the explosion of press attention, I asked for an administrative inquest that would focus especially on the communication of confidential administrative documents with the intent to damage my reputation. This request almost backfired on me. Several weeks after I had placed it, the president of the university (a biologist and member of the national committee on ethics) and a local scientific celebrity (a biochemist, member of the Academy of Sciences, very close to Jaubert) criticized my attitude in the *Caulerpa* affair during a local biologists' meeting. In the face of these remarks, I defended myself and convinced the committee of university biologists to remain neutral. However, the administrative inquest dragged on. A committee was created. More than one year later, it provisionally dismissed the case. The committee ruled it lacked authority to judge on a litigation they described as "purely personal and separable from university activity." They drew no definitive conclusion on the substance of the case, and, despite my written appeals, I was never permitted to address this committee.

1993–94: THE MUSEUM PRODUCES PRO-*CAULERPA* FILMS

Having failed to win their case by the judicial route, the defenders of *Caulerpa* again engaged the public directly. Having observed that the *Caulerpa* videos had swayed some undecided individuals, the Museum directorate decided to invest in a more intensive production of films on the theme of "*Caulerpa,* haven of life."

The Oceanographic Museum financed, to the tune of more than a million francs, a system allowing round-the-clock filming of the *Caulerpa* fields beneath their building. Two stationary cameras, furnished with powerful lighting, were attached under the Museum in twenty to thirty meters of water. The images, transmitted by underwater cables, were projected live on a giant screen in the public lecture room of the Museum. A collection of the most beautiful scenes of life among the *Caulerpa* was selected from hundreds of hours of recordings. The montage gave the impression of a teeming sea. By day could be seen several octopuses and beautiful schools of fishes. At night, many fishes, attracted by the lights, crossed the field of vision of the cameras. This is the "lamparo" effect (the lamparo is a powerful light used to attract fishes at night; this fishing technique is prohibited in many Mediterranean regions because it is so effective and destructive). This new video, presented to the press and to many government authorities, did not garner the expected media interest. The artifice was all the more obvious because scientific research was beginning to prove that the numbers and sizes of certain fish species tended to drop in sites invaded by *Caulerpa* and monitored assiduously for two years.

At the end of the spring of 1994, Doumenge's cameras filmed a phenomenon that he did not recognize: the seasonal proliferation of filamentous brown algae. The sporadic growth of Mediterranean algae, studied for many years by Italian phycologists, took place above the *Caulerpa.* Doumenge judged it an opportune time to announce pathetically to the local television and press the sad end of the killer alga that was so environmentally useful.[21] Several weeks later the filamentous algae disappeared, leaving the lush, verdant *Caulerpa* prairie in the heart of its regeneration cycle. In the autumn, everything returned to normal: *Caulerpa* again covered nearly 100 percent of the field of vision of the cameras. But it did not matter! The films heralding the end of *Caulerpa* had been distributed to journalists who echoed them: "*Caulerpa:* the devourer devoured," "The parasite parasitized," "Considered by some to be the 'danger of the century,' *Caulerpa taxifolia* is in turn threatened by another alga."[22]

JANUARY 1994: THE FAILURE OF THE FIRST SCIENTIFIC COMMUNICATION
OF THE DEFENDERS OF *CAULERPA*

Reacting to these strenuous publicity efforts, scientific researchers presented an opposing argument that was beginning to convince people; *Caulerpa* was never the object of a scientific controversy because Doumenge and Jaubert had never written any scientific publication. To this point, the debate was solely in the media. The advocates of *Caulerpa* were thus forced to address the problem in the traditional way, by submitting their ideas to the unbiased judgment of scientific journals.

By this time, Doumenge and Jaubert had in fact changed the subjects of their research. Having abandoned the idea of cultivating the Persian Gulf corals at the Museum in order to save them, they disclosed a new project to the press—that of using their coral culture technique to fill in the fissures generated by the nuclear explosions in the Mururoa Atoll. Jaubert even created a private corporation to commercialize the application of his invention.[23] With such a project, research on *Caulerpa* would cease to be a top priority.

The first international seminar on *Caulerpa taxifolia*, planned for January 1994 in Nice, was to give the various scientific teams the opportunity to report the first results of the studies financed by the European Union. The main organizer of the colloquium, Boudouresque, received a paper submitted by the Scientific Center of Monaco and signed by Jaubert. The title referred to a research program on the effects of light, temperature, and nutrients on the growth of the alga.[24] The submitted

198 text was far from conforming to scientific norms; presented with polemic connotations, it reported no results, but stated intended research. Not wanting to oppose this step by the researchers based in Monaco, Boudouresque contacted Prince Rainier directly to express his satisfaction at seeing the Prince's research institute involving itself in the *Caulerpa* problem. He explained he hoped for a more scientific style. The Prince responded quickly; the Scientific Center of Monaco had no intention of getting involved in a polemic over this matter. He specified that Jaubert "would not stray from the strictly scientific domain."[25] Jaubert in turn sent a second summary that better conformed to scientific standards, and their new text, signed by three other researchers, was accepted.[26]

During the colloquium, one of the co-authors presented the results in a correct format, but these results dealt with only certain requirements of the alga, that is to say with only part of what was announced in the title. Jaubert attended the presentation of his work, then deposited a text on tables where abstracts of the other colloquium communications were made available to scientists, invitees, and journalists. This text was not the one that had been accepted and presented, but a new version closer to the one that had been rejected. Evidently, this maneuver was aimed not at convincing the scientific community, but at influencing reporters.

MAY–DECEMBER 1994: THE PSEUDO-SCIENTIFIC PUBLICATION

Five months later, in May 1994, the same Museum strategists rekindled fires by again taking up one of the first Doumenge sophisms on the natural origin of *Caulerpa taxifolia* in the Mediterranean. According to Doumenge, *Caulerpa taxifolia* had to be considered a relictual species that had survived the closing off of the Mediterranean more than five million years ago (it had previously been open to the Indian Ocean) and its subsequent drying. This thesis had not been announced in scientific circles until then, but only to the press. A scientist was now going to carry the torch—the Sicilian Professor Giuseppe Giaccone. With one of his assistants, he affiliated with members of the laboratory of the Scientific Center of Monaco to undertake research on *Caulerpa taxifolia* financed particularly by the Principality.

For the meeting of the Italian Commission on Marine Biology in 1994, I was invited to participate in a roundtable on *Caulerpa taxifolia*, as were Doumenge and Giaccone. Doumenge showed his documentaries on the fishes filmed by his underwater cameras, while admitting their preliminary nature and that they did not have real scientific value. Giaccone participated in French, recalling the history of communication between the Indian Ocean, the Red Sea, and the Mediterranean. According to him, the messinian crisis that took place five million years ago, the geological upheavals of the Plio-Pleistocene, the canals linking the Mediterranean to the Red Sea dug successively by Ramses II (1300 B.C.), by the Caliph Al Mansur (in the year 800), and most recently by Ferdinand de Lesseps (1869) explained the presence of numerous tropical algae in the Mediterranean. This long historic presentation, illustrated by well-founded examples of migrations of species from the Red Sea to the eastern basin of the Mediterranean, ended with a small, ambiguous sentence: "The same mechanism seems to us to be followed by *Caulerpa taxifolia* in the western Mediterranean, although it does not seem legitimate to speak of migration for this species, if one considers that it was a question of an accidental introduction."

He then referred to the concentration of toxins in *Caulerpa taxifolia*, which, according to him, were negligible and comparable to those in other common algae of the Mediterranean. He claimed to have based his reasoning on the work of scientists who had found practically no trace of the main toxin (caulerpenyne) in *Caulerpa taxifolia*.[27] However, the cited publication made no reference to the concentration of this toxin; on the contrary, the same authors had, in a previous publication, emphasized the presence of a great quantity of caulerpenyne in *Caulerpa taxifolia*![28] Giaccone used tactics well known to sociologists of science: bury the reader under an avalanche of precise facts to try to validate another idea that is poorly supported.[29]

I did not have much trouble deciphering Giaccone's errors. Naively,

200   I thought the confrontation would stop there, and I was satisfied with having demonstrated the anomalies before an audience of specialists who would not be duped by improper reasoning. Giaccone had expressed a personal opinion in a debate among scientists, and the text,[30] the basis for the oral presentation, had not been reviewed.[31] In fact, the text was written for the media. Copied by the Museum offices, it was to be distributed to the press with a handwritten headline: "Alghero (Sardinia), May 24–28, 1994. 25th Congress organized by the Italian Commission on Marine Biology." The fuzziness deliberately maintained by this heading (was it an oral or written communication, published or not published, approved by referees?) was not noted by the reporters. Some fell into the trap set with this pseudo-scientific publication by a respected society of biologists from a neighboring country. Thus one could read in *Le Monde* that the alga, of natural origin, contained almost no toxins.[32] A local newspaper took this text as proof that the alga was innocuous: "The harmless *Caulerpa*. At the twenty-fifth Italian congress of marine biology, a communication concluded that the alga is harmless and in no way a 'killer.'"[33]

Fortified by this foreign aid, Doumenge and Jaubert reignited the polemic. The former reentered the fray with a thundering declaration in *Le Midi Libre*,[34] newspaper of Montpellier, where he had begun his university career:

Reporter: Caulerpa?
Doumenge: I eat it in doughnuts with the Prince!
Reporter: The scientific alarmists?
Doumenge: You will soon be able to read my work ridiculing these opportunists; it is entitled *Caulerpa: The myth and the reality.*"

Four years later, his book has yet to appear. Was the Prince used as a guarantee of the absence of toxins in the alga? Did he really taste the *Caulerpa* from the Mediterranean, or had he eaten *Caulerpa lentillifera*, the Philippine "poorman's caviar"? The episode of the "Prince's doughnuts" was to amuse many a scientist.

1. (TOP) Vertical rocky cliffs covered by *Caulerpa taxifolia* (Cap Martin).

2. (BOTTOM) Rocky substrates covered by *Caulerpa taxifolia* (Cap Martin, 26 meters with gorgonians).

3. (TOP) Develop-
ment of rhizoids
from 1 mm cutting
of *Caulerpa taxifolia*
(1 week).

4. (BOTTOM) From
a cutting, the alga
colonizes a sandy
bottom by elongat-
ing its horizontal
axes (stolons) at
the island of Hvar,
Croatia, in the
Adriatic Sea.

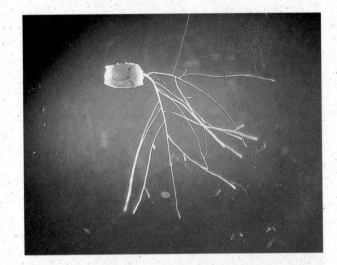

5. (TOP) Boulders entirely covered by *Caulerpa taxifolia.*

6. (BOTTOM) Area of sand covered by *Caulerpa taxifolia* (Cap Martin).

7. (TOP) Fronds of *Caulerpa taxifolia*
before and after having been sucked
by a sea slug.

8. (BOTTOM) A tropical sea slug,
Elysia subornata, eating *Caulerpa
taxifolia*. This species is a hoped-for
biological control agent.

In 1995, the defenders of *Caulerpa* introduced a more technical argument to the battle. To understand what was at stake, we have to review briefly the hypotheses successively introduced by Doumenge. In order to prove the natural presence of *Caulerpa taxifolia* in the Mediterranean, he stated in 1991 that the alga proliferating under the Museum was not *Caulerpa taxifolia* but a common species of *Caulerpa* that had always existed in the Mediterranean. In 1992, he advanced the hypothesis of a "dormant" species in the Mediterranean that had reappeared because of the combined effects of increased temperature and pollution. Then he affirmed, in a letter addressed to the interregional director of Maritime Affairs (dated February 21, 1992), that the species *taxifolia* was identical to *Caulerpa mexicana*, a species that had been recorded in several Mediterranean locations (Israel, Lebanon, and Syria).[35] Thus "*Caulerpa taxifolia-mexicana*" had always existed in the Mediterranean.

To cut to the quick of these unscientific hypotheses, I drafted an article on the two species. I assumed this clarification, presented at the first seminar on *Caulerpa taxifolia*,[36] would definitively end a fruitless polemic. But in June 1994, this thesis was taken up again by Jaubert, who distributed to reporters three texts mentioning "the flood hypothesis" (the messinian crisis).[37] Less prudent than Giaccone, he stated confidently that the colonization of the Mediterranean by many tropical species is only a natural episode in the reconstruction of the ancient, pre-messinian tropical habitat that had once reigned in the Mediterranean. Some species must have survived the cataclysm and others are constantly migrating from the Red Sea towards the Mediterranean. In this scenario resides, according to Jaubert, the whole explanation: *Caulerpa taxifolia* is just part of the group. As proof, he said that *Caulerpa taxifolia* was long known in the Mediterranean as *Caulerpa mexicana*. *Taxifolia* and *mexicana* were thus synonymous.

This hypothesis was the focus in 1995 of efforts by Jaubert, with the help of Giaccone. It was the latter who identified the *Caulerpa taxifolia* found at Messina by a Palermo researcher as *Caulerpa mexicana*.[38] We were thenceforth to believe that *mexicana*, previously restricted to the eastern basin of the Mediterranean between Tel Aviv and Beirut, had just arrived in Sicily and had transformed itself into *taxifolia*. This "transmutation" hypothesis was first published in the Italian press.[39] Once again the news media were used to contradict a scientific publication. The media circus was heated in Sicily, where the clarifications of the researchers from Palermo and Catania followed one another. In 1995, Giaccone presented his theory at the 26th congress of the Italian Society of Marine Biology. He affirmed that "the alga from the Straits of Messina is not *Caulerpa taxifolia*, but the less hated *Caulerpa mexicana*." As the only justification for this claim, his abstract was illustrated by a figure showing the real differences between the two species, a figure taken from the publication of another author who had compared two samples collected in the Indian Ocean.

Jaubert furthered the justification of the equivalence of *Caulerpa mexicana* and *Caulerpa taxifolia*. While cultivating some samples of *Caulerpa taxifolia* found at Messina (of which certain individuals had short, wide pinnules), he observed that it transformed into a more typical form of *Caulerpa taxifolia*, similar to those growing in front of the Museum (with narrow, long pinnules). He named the form growing at Messina "*Caulerpa mexicana*" and felt that *mexicana* could transform into *taxifolia*. In a work dedicated to the activities of the Principality of Monaco for the protection of nature, he presented the results of his research and, for the first time, called the transformation of the *mexicana* from Sicily into *taxifolia* from Monaco a "metamorphosis."[40]

In order to validate his "discovery," he had to publish it in a prestigious journal. But this would require buttressing the "metamorphosis" hypothesis with more scientific data, not just simple photographs that, moreover, did not show the other criteria for distinguishing between the two species, let alone genetic evidence. Jaubert and Giaccone submitted an article to a scientific journal, the *Comptes Rendus* of the French Academy of Sciences. The article was accompanied by photographs, but one of the traits separating the two species was not depicted. The article explained that the analysis of the toxins had been assigned to two different chemists, Vincenzo Amico of the University of Catania and Philippe Amade of INSERM, based at the University of Nice–Sophia Antipolis. Jaubert and Giaccone had provided them with samples of *Caulerpa* from Messina, from the Côte d'Azur, and from various stages of what they called the "metamorphosis." This article was subsequently contested.[41] The whole story was associated with many precise data on the global warming[42] of the Mediterranean. The *Comptes Rendus* article was translated into English by an Australian postdoctoral student.[43]

The eminent academicians took the theory seriously. In this august assembly, the biologists on the editorial committee are, with one exception, all "reductionists," which is to say they are researchers who study living organisms only at the molecular level. With little interest in the classification of algae or marine ecology, they could not know the context of this complicated story as long as no academic botanist or corresponding oceanographer was consulted. By the same token, they heard from none of the critics of the thesis that was being presented. Despite the reservations of a specialist in algal genetics, the academicians accepted the article after several revisions. Knowing that they were publishing a scoop, the Academy announced the publication with a press release, "because of its ecological impact of the greatest importance, we are convinced that this article will capture the attention of your readers." The secretary-general of the editorial board even stated that the article "will not fail to rekindle the polemic on the origin of this tropical alga on our shores."[44] On this point, the academicians were not mistaken!

The *Comptes Rendus* publication provided the first scientific support for the idea that the alga did not come from Monaco. Jaubert distributed his article to the press, accompanied by a press release.[45] Although the media controversy had subsided, the reporters again made the rounds of the protagonists in the affair. Some showed a blind confidence in the theory that had been swallowed by the French Academy:

"It appears difficult to impugn the scientific integrity of such a piece of research, of which the outcome is presented under the auspices of one of the most venerable French institutions."[46] "An hypothesis that is *a priori* trustworthy because it emanates from the Academy of Sciences has just made some recent theories on the subject look very bad."[47]

Knowing all too well that a simple counter-publication would not suffice to reestablish the truth, I suggested to the *Comptes Rendus* editorial board that the journal itself take on the task in an editorial. Three weeks later, the permanent secretary of the Academy, François Gros, suggested an intermediate solution that did not satisfy me. I would submit a counter-publication that would be quickly considered by the Academy. It was hard for me to understand the lack of any direct reaction by the editors. With their vague promises of an attentive examination of possible earlier errors, and with wide publicity assured for a counter-article, I reluctantly drafted a response on the origin of the alga and denounced the errors and egregious anomalies of the "metamorphosis" article. Boudouresque joined me in this reply. We sent our article to the Academy on March 8, 1996, along with a letter recalling once again what seemed to us to be the proper editorial action in such an affair. It seemed to us that the Academy should have done more than simply grant us a right to publish a reply. Our counter-article was accepted after we were forced to remove several items relating to the initial data. The semantics of science avoid polemical terms. Thus, the terms "modification of number of examined samples" and "selective use of data" were preferred to more pointed words.[48]

In scientific circles, this affair inspired several satires, of which the following was the most widespread. On the "transmutation" of *taxifolia* into *mexicana:* "You take a goat, you call it a sheep. Then you observe that it still looks like a goat. And you conclude: the goat and the sheep are actually one and the same species." On the "metamorphosis" of a *taxifolia* with wide pinnules into a *taxifolia* with narrow pinnules: "When you take your geranium from your balcony to your salon, the leaves become larger and more widely spaced. Does that mean it has 'metamorphosed?'" On the hypothesis of the migration of a *Caulerpa* that came from the Red Sea because of a warming of the Mediterranean: "A lion is discovered fifty meters from the door of the Vincennes zoo. The director of the zoo, or a scientist whom he sponsors in his facility, tries to show that the lion is present in Africa, that it could have arrived naturally in Vincennes by way of the Middle East. That global warming is the explanation for this migration (and it is true that the warmest eight years of the last 140 years were all after 1980). That the lion was present in southern Europe during geological times (the Quaternary), which is also true. In short, that there is an alternative hypothesis to a zoo escape, and even that there is a 90 percent probability that this alternative hypothesis is the correct one. What do you think of that? Ancillary question: Would the *Comptes Rendus de l'Académie des Sciences* publish such fantasies?"

Thomas Belsher, phycologist at IFREMER who has worked on *Caulerpa* since 1992, sent some reporters his severe criticisms accompanied by a humorous postscript pointing to a missing reference in the article of the Academy of Sciences: "Copperfield, David, 1995. Feats and magic in the western Mediterranean. The special case of the metamorphosis in the vegetable kingdom. Edit. Nostradamus, p. 327." Imagine his surprise when a reporter insisted on obtaining a copy of this reference! The reporter for the British science magazine *New Scientist* also did not fail to employ humor in concluding his inquest on the affair of the "metamorphosis" by this appeal to "return to original sources": "The Oceanographic Museum of Monaco is soon going to complain that the lessepsian migrant has climbed up the sea water system and invaded the aquaria!"[49]

If the title of Jaubert's article suggested only one alternative to the hypothesis of inadvertent introduction of the alga from the Museum aquaria, he did not hesitate to be much more definitive in his subsequent statements: "Even if one cannot speak of absolute certainty, it is 90 percent probable that the explanation lies in that direction, asserted Professor Jaubert."[50] His other statements tended to establish his independence from the Oceanographic Museum of Monaco: "Rejecting the existence of any financial or administrative connection with the Oceanographic Museum of Monaco, Jean Jaubert feels that Alexandre Meinesz and the scientists who support him are part of 'an enormous lobby that is trying to get money.'"[51] In *Sciences et Avenir,* "Professor Jean Jaubert, director of the European Oceanographic Observatory in Monaco, who is anxious to establish that he has no financial connection with the Oceanographic Museum of the principality."[52] And the *New Scientist:* "Jaubert argues that he does not have a vested interest in proving that the algae proliferating in the northwestern Mediterranean did not come from the Monaco museum, as his institute is a separate organisation."[53] Jaubert systematically emphasized his total independence.[54] His press release stated: "To maintain its independence as well as objectivity and efficiency, the European Oceanographic Observatory has kept its distance from matters (committees and programs) that were clearly too influenced by the promoters of alarmist theses." A reporter received another explanation: "The document was drafted under the aegis of the European Oceanographic Observatory, of which the laboratories are located—by sheer coincidence, Professor Jaubert assures us—within the walls of the Oceanographic Museum of Monaco."[55] Another "coincidence" is not surprising: the study straining to prove the "natural" migration of the alga from the south to the north of the Mediterranean, which contradicts the Monacan origin of the alga and its dissemination by yachts, was financed jointly by the Principality of Monaco and an international magazine that specializes in advertisements for leisure craft.[56]

By contrast, we were suspected of partiality. We were supposedly concerned above all with defending a new European program of studies that was under discussion, with the ultimate goal of feeding the coffers of our laboratories. A press agency headline stated "New polemic between scientists based on renewal of European financing."[57] The hypothesis of a rivalry between the laboratories of Nice and Monaco resurfaced: "Why does one get the regrettable impression that *Caulerpa* is poisoning most of all the climate between the laboratories of Nice and

Monaco? It is that we are also dealing with matters of the coin of the realm."[58] "*Caulerpa taxifolia:* war of the laboratories."[59]

In fact, this *n*th revisitation of the scientific/media debate forced us once again to deal with the algal origin, a colossal waste of time. In 1996, the really important question for us was not to know if the alga had been introduced to the Mediterranean by the Museum or had arrived from the Red Sea, but to assess the risk posed by the alga to Mediterranean ecosystems. But the thesis published by the Monacans suggested a natural element of the invasion. We had to react. For that reason, in an article sent to the Academy of Sciences, we amassed all the knowledge bearing on the arrival of the alga via the aquarium network, before doing a normal scientific critique of the "metamorphosis" of *Caulerpa.*

As for origins, the geography of the invasion was eloquent. At the end of 1995, 90 percent of the Mediterranean area known to be invaded by the alga was still found on either side of Monaco (between Menton and Nice). At this time, *Caulerpa taxifolia* (which is not a metamorphosed *mexicana*) had just arrived at Messina and in the Adriatic, where it did not yet cover one hectare. In the western Mediterranean, the invasion of *Caulerpa taxifolia* was rapid: from 1984 to the beginning of 1996 nearly forty sites were invaded; the total area of more or less covered zones was more than a thousand hectares. During the same period, on the other side of the Mediterranean (Syria, Lebanon, and Israel), *Caulerpa mexicana* (the real one) remained rare. From 1939, the date it was first observed, no assessment of the area covered by it has been conducted. In this zone of the eastern Mediterranean, the water during the cold season is 4–5°C warmer than at Monaco. Those who hold that the 0.1 to 0.5° general increase in Mediterranean water temperature is responsible for the invasion of the northwestern coasts by a *Caulerpa* coming from the south have not explained why it did not first invade the warm waters of the eastern Mediterranean! This "anomaly" escaped the editor.

For the scientists based at the Oceanographic Museum of Monaco, only molecular genetics could truly determine the identity of the species of *Caulerpa*. This approach would require dragging the referees and protagonists into long, laborious, and costly scientific research. In its place, we have advocated simply pinching off and identifying samples of *Caulerpa* from several sites, including a site with water cold by Mediterranean standards (between 10 and 15°C). Only the aquarium clone of *Caulerpa taxifolia* is known to be resistant to such low temperatures. The results of this experiment are illustrative: the *Caulerpa* found in the more southerly regions of the western Mediterranean (Split in the Adriatic, Messina in Sicily, and Majorca in the Balearics) have all survived after several weeks of growth in the cold; by contrast, samples of *Caulerpa taxifolia* and the true *mexicana* from the Caribbean have lasted only a few days under 17°C. A little common sense goes a long way.

Our retort appeared in the *Comptes Rendus* in midsummer.[60] Contrary to the lavish Academy press release that accompanied the initial article, our paper had no particular support from the academicians. We therefore had to publicize it ourselves. A bit offended by this publicity, three academicians sought to justify the publication of two articles in a row on the same topic as ordinary conflicting scientific accounts.[61]

The Ministry of the Environment asked the Academy of Sciences for an assessment of the two articles. Lists of international experts were assembled, and the Academy drew up a budget to pay for its collaboration. The assessment was transformed into an international seminar on all the problems related to *Caulerpa taxifolia*, finally held in March 1997 (see chapter 6).

In the scientific domain, it is not considered good form to publicize dysfunctions of the system. Scandals damage the image of science in general.[62] This institutional reflex partially explains how the environment can become hostage to a scientific war. The adherents of the hypothesis of a species transformed under the influence of global warming were bound to have a hard time holding their last line of defense. At most they could continue to retard the public recognition of the growing ecological threat to the Mediterranean.

Biological Control: The Last Hope?

In June 1992, I received an encouraging letter from an American re-searcher in Melbourne, Florida. Kerry Clark had learned about the *Caulerpa* invasion from a television show in his country. An ecologist specializing in small, tropical marine slugs (sacoglossans, also known as ascoglossans, in the Mollusca), he had conducted research on particu-lar species that fed on *Caulerpa*. He suggested using these slugs to re-duce the growth and spread of *Caulerpa taxifolia*. This is the prin-ciple of biological control, a technology commonly used in agriculture to fight "naturally" against invasive, undesirable species. For instance, in the southeastern United States, South American aquatic alligator-weed *(Alternantheres phyloxeroides)* became a severe problem, choking many water bodies. But it was largely controlled, beginning in the 1960s, by an introduced flea beetle *(Agasicles hygrophila)* from its native range.[63]

After I received his publications on the subject, I learned of a small group of slugs that fed exclusively on certain *Caulerpa* species. But, un-like the tropical *Caulerpa* individuals, which formed small clumps scat-tered on the sea floor, the Mediterranean clone of *Caulerpa taxifolia* extended over very large areas, with elevated densities and a rapid rate of growth. I was thus a bit skeptical that biological control using slugs would be effective.

However, a succession of fortuitous circumstances allowed us to test Dr. Clark's idea. In the context of the research program undertaken by the European Union, our laboratory was assigned to study the different behavior of Mediterranean *Caulerpa taxifolia* and tropical *Caulerpa taxifolia*. The Mediterranean clone handled temperatures of 13°C very well during the winter months, while in all tropical regions, the alga never extends to zones where the water temperature falls below 20°C. We decided to verify the temperature requirements of *Caulerpa* in aquaria, in growth chambers with controlled temperatures. We asked a phycologist in Guadeloupe to send us a living sample of tropical *taxi-*

210   *folia.* In twenty years of harvesting algae in the Caribbean Sea, this species was so rare in the region that she had never noticed it. By luck, she discovered a small colony on some rocks near the surface, but how could a living clump of *Caulerpa taxifolia* be shipped? On the first try, the algae, conditioned either in a humid bag with an oxygen-enriched atmosphere or inside a flask containing seawater, were shipped in an ice chest intended to keep them at the right temperature. But these *Caulerpa* all arrived in Nice dead from the cold, probably because the layover between Paris and Nice was too long. A new ice chest containing *Caulerpa* was given to the crew of a flight from Guadeloupe to Paris. Kept warm in the cabin, it was supposed to be met in Paris by one of my students. But because of an unusual storm, the airport at Roissy was closed and the plane from the Antilles was rerouted to Bordeaux. When we were able to pick up the ice chest, the *Caulerpa* were again dead from the cold. The only solution was to send one of my collaborators whose parents lived in Martinique into the field. Finding no *Caulerpa taxifolia* at Martinique, he went to Guadeloupe—while his mother meantime discovered the alga on Martinique. Finally, a substantial quantity of *Caulerpa* from these two islands reached Nice in perfect condition. An aquarium of one square meter, heated to 25°C, had been prepared for their cultivation.

We drew from the stock of these tropical *Caulerpa* to undertake a large number of comparative experiments with the Mediterranean *Caulerpa.* The latter were in fact very different from those of the Antilles. They could not be distinguished morphologically, but the Mediterranean *taxifolia* always grew much more rapidly than the tropical individuals. The Mediterranean individuals tolerated the cold very well and even survived three months at 10°C, whereas below 20°C, the tropical individuals died after several days. These experiments demonstrated the existence of genetically based differences. One day, the student assigned to maintain the aquaria came to me in deep distress: the tropical *Caulerpa* were dying. According to him, a small green slug was eating them—Clark's slug! We quickly discovered a dozen molluscs in the aquarium, belonging to two different species that must have been stow-

aways from the Antilles. In a new aquarium, we placed the slugs with a
clone of *Caulerpa taxifolia* from the Mediterranean. In several days,
they ate it all and laid dozens of strings of hundreds of eggs each in a
frenzy. They feasted on the "killer alga" and reproduced in exponential
fashion. It was too good to be true.

I gathered the laboratory team to define a research plan that would
be conducted in great secrecy; we would unveil our study only when a
scientific communication could be presented. In April 1994, an American student from the University of Rhode Island, in training in our laboratory, was assigned to conduct the first experiments and determine the
quantities of *Caulerpa* ingested by the two slug species in the temperature conditions of the Mediterranean: between 12 and 26°C. Other students took up the baton. We quickly succeeded in growing populations
of several hundred slugs of each species that every day devoured bucket-
fuls of *Caulerpa taxifolia.* At all seasons and whatever the weather, they
had to be fed with *Caulerpa.* We informed Clark of our discovery and
asked him to confirm our determination of the two slug species. They
had the pretty names of *Elysia subornata* and *Oxynoe azuropunctata.*

Observing the slugs and studying a voluminous literature taught us
the extraordinary biology of these molluscs. First of all, we noticed that
we had, by luck, found the most interesting species. Our slugs showed a
preference for a small group of *Caulerpa* species, those with axes and
pinnules of small diameter, as in *Caulerpa taxifolia. Elysia* and *Oxynoe*
envelop parts of the alga with their mouth, than apply a sharp rasp, in
the form of a row of teeth, called a "radula." Through the opening that
they make in this manner, the slugs suck out the cellular sap of the
Caulerpa (its cytoplasm) thanks to a very muscular pharynx. The strong
viscosity of the sap forces them to renew the operation often: the slug
repeatedly glides several millimeters to establish a new "foraging" site.
In several hours, a frond of *taxifolia* can thus be pierced several dozen
times and the cytoplasm sucked out. The frond, drained and dotted with
whitish haloes, loses its rigidity, hangs sadly, and dies without regrowth.
The "vampires" of *Caulerpa taxifolia* are not able to eat other algae;
either the algal wall is too thick, or the diameter of the axes is not suited

to the sizes of their mouths, or the anatomy of the algae is such that they cannot be ingested by aspiration. The marine angiosperms (species of *Posidonia, Cymodocea, Zostera*) and the great majority of algae are multicellular; they cannot be sucked out. But *Caulerpa*, with its single giant cell, can be pierced and its contents ingested by the slugs.

This adaptation to one type of food is not the only peculiarity of our slugs. In tropical waters, *Caulerpa taxifolia* or other *Caulerpa* species that have axes with small diameters are not very common. After consuming a clump of *Caulerpa*, the slugs then have to traverse long distances for several days without eating anything before finding another clump of *Caulerpa*. How have these slugs adapted to long periods of famine? By an astonishing mechanism. While swallowing the cell sap of the *Caulerpa*, they ingest the chloroplasts of the alga. These organelles, which contain chlorophyll, are the little factories that use the energy of the sun and carbon dioxide gas to construct the vital nutrients for the plants. As a rule, all animals that eat plants digest the chloroplasts along with the rest of the cellular contents. Not *Oxynoe* and *Elysia*, which preserve them and store them near the surface of their skins. After devouring *Caulerpa*, the slugs set out to find a new clump and activate their solar battery (consisting of the store of chloroplasts) by stretching their skin as much as possible to capture the rays of the sun. Using their bodies as leaves, they then function as plants! Scientists who have studied the phenomenon have shown that, when well lit (daylight is sufficient), these slugs can survive and move for several dozen days without eating. Without daylight or food, they die in several dozen hours.

In their hunt for *Caulerpa*, the slugs have to cross hostile terrain where dozens of predators can devour them. Loaded with chloroplasts, they take on the green color of *Caulerpa*, which makes them highly visible the moment they leave the clumps of green algae. However, they are never eaten. This is because, while sucking the cell sap of *Caulerpa*, they swallow the algal toxins, which they transform into other toxins and then store. They are thus poisonous to their enemies. Tropical animals know this and avoid them. Aquarium experiments with Mediterranean fishes, starved for several days for the trial, show that they attack the prey, but then quickly spit out the slugs, never bothering them again. The slugs even protect themselves from bites; as soon as they are disturbed, they exude a white cloud. Untouchable, because repulsive, the slugs can then continue on their way.

These incredible slug behaviors combine with an equally interesting reproductive strategy. In the tropics, *Elysia* and *Oxynoe* mass in great numbers at a clump of *Caulerpa;* they reproduce quickly and devour the *Caulerpa* in short order. Then they scatter, each in search of another clump. Before parting on this adventure, the slugs, which are hermaphrodites, carry with them a little bag of sperm acquired during copulation with another slug. The sperm, sometimes conserved alive for several months, are used little by little for self-fertilization. As soon as a slug has found a patch of *Caulerpa*, it lays a thousand eggs per week; it is quickly surrounded by thousands of descendants.

Another unusual trait of our two slugs made us think even more about our good fortune in finding them. In the majority of sacoglossans, the larvae float and are dispersed by the currents (the planktonic lifestyle). The dissemination of the larvae is by chance; one among thousands will not be eaten by a predator and will land on a clump of *Caulerpa*. This is also the case for the few species of Mediterranean sacoglossans that eat *Caulerpa prolifera*. The two tropical slugs that we studied were different. Their larvae have yolk sacs that nourish them longer; they do not leave the egg until after they have metamorphosed into tiny slugs that are soon able to quench their desire for their exclusive food: *taxifolia*. A single slug can therefore engender, several months later in the same place, several hundred thousand descendants. Thanks to this peculiarity of their life cycle, one could hope to find millions of slugs in the *Caulerpa* fields of the Mediterranean. And there would be assurance that, after they had eaten everything, they would survive only a few days thanks to their battery of chloroplasts, without being able to eat other marine plants.

But how do the tropical *Caulerpa* resist vampires so highly organized and voracious? By their sexual reproduction. Regularly, in fact, tropical *Caulerpa* individuals liberate gametes of both sexes that unite to form eggs (zygotes) that attach to the bottom and reconstitute new colonies far from slug-infested zones. The young *Caulerpa* clumps can thus grow for some length of time far from their specialized predators. On the other hand, no sexual reproduction has been observed in the clone of *Caulerpa taxifolia* introduced to the Mediterranean. Only sperm seem to be produced several days per year: the Mediterranean *taxifolia* are thus all males. Lacking the ability to multiply rapidly and effectively by sexual reproduction, these *Caulerpa* ought theoretically to be able to be eradicated by the vampire slugs. Finally, our research showed that, below 20°C, the tropical slugs are less voracious. Between 14 and 16°C, they remain immobile. At this temperature or below, they die of the cold.

The biological characteristics of *Caulerpa taxifolia* and of their specialized predatory molluscs allowed us to envision an eradication of the alga by biological control. Our experiments led us to three conclusions.

First, several slugs could not finish off all the *Caulerpa*. They could not be released until the end of the spring, when the water temperature of the Mediterranean reached 18°C. They would also have to be released on all the sites and we would have to rely on their formidable reproductive potential to generate enough of them to be able to eat a lot of the *Caulerpa* before the arrival of winter, which would inevitably kill them (13°C on average in the northwest Mediterranean from January to March). The number of slugs to release would be determined by the area to be treated. If *Caulerpa* remained at the end of the season, the operation would have to be renewed the following year.

Second, two natural "guarantees" theoretically limited the environmental risks of experimentally introducing a new species into the sea. For one thing, the slugs would not survive long in the absence of *Caulerpa* (after having eaten all of them); for another, they would be condemned to die of cold when the winter came. Between their introduction in May and their disappearance in December, only three or four generations could grow. The low levels of gene flow combined with the mortality in winter ought to limit the evolution of resistance to the cold, at least in the northwest Mediterranean.

But, on the other hand, the deep zone where *Caulerpa* grows (beyond thirty meters) is always temperate or cold, whatever the season (below 18°C). At these depths, the alga would have a refuge from the slugs. Total eradication thus appeared impossible.

The speed of reproduction by the slugs in the ambient conditions of nature is difficult to predict. We were reduced to assessments based on our laboratory observations. A real schoolchild's calculation—if a slug lays 1,000 eggs every ten days and if, after fifteen days, at least 200 larvae hatch from each clutch of eggs yielding little slugs that grow to adulthood after a month and a half and lay their own eggs in turn, how many square meters of *Caulerpa* will be eaten by the slugs in five months at a rate of three fronds per adult slug per day on a carpet of 5,000 fronds per square meter? According to our estimates, the key to the battle would be the last month, after the growth of the last generation before winter, when the slugs are sufficiently numerous to do real damage. Depositing a clutch in May ought to lead to a population of several million slugs come November. But this multiplicative progression is completely theoretical. Is it realistic? We would have to try it in the sea. Perhaps the slugs would be devoured in the larval stage before they became toxic. Perhaps the expected explosion of the slug population would not take place.

In December 1994, we presented our slugs to the second international colloquium on *Caulerpa taxifolia* at Barcelona. Of an assembled 120 researchers, a third encouraged us in our course of action, a third remained skeptical, and the last third were hostile to the idea. These latter individuals thought that we would risk introducing into the Mediterranean a new scourge more dangerous than *taxifolia*. They recalled the many unfortunate examples of biological control in which species intended to reduce a proliferating invasive species instead attacked a different, native species. For example, the Florida and Central American carnivorous rosy wolf snail *(Euglandina rosea)*, introduced to the Hawaiian islands, Tahiti, and many other places in a futile attempt to control the imported giant African snail *(Achatina fulica)*, has instead caused the extinction of many native species of harmless tree snails.[64] However, in other instances there were real successes with this method of control. But it is true that biological control has never been used in the sea, because the proliferation of harmful introduced species there is a rarer and often ephemeral event. Also, I was determined on the one hand to begin the first tests in the sea only under the control of an international commission that could examine all the information on the subject, and on the other hand not to distribute slugs to other laboratories that might have fewer scruples.

The only French journalist present at Barcelona belatedly published this information in a scientific magazine for young people.[65] The news then spread to other media. To the reporters who contacted us, we gave our publication, and we were careful to explain that this project was only an idea at this point. A promising idea, but until the aquarium tests were confirmed in the sea, it was important to be prudent.

One year after our scientific communication, only the chief of the mission for the Ministry of the Environment (Manche), seduced by the idea, had made an effort to help us. But he did not succeed in convincing the decision-makers, if even to propose a fellowship to a thesis student or to ameliorate the mediocre conditions in which we reared the slugs.

Along with the progress in our studies of the slugs and the proliferation of the alga, a new cause for anxiety arose. We realized that if we used biological control too late, the method could be rendered ineffective because the continued expansion of *Caulerpa* diminished the reasonable prospects of an effective control. There would be too much alga even for the hungriest of slugs. For the project to be successful, the slugs would have to be released simultaneously at all contaminated sites. Now, at the end of 1995, the inventory of contaminated sites had become very uncertain. A complete grasp of the overall extension of *Caulerpa taxifolia* eluded us, in France, in Italy, in Croatia, and even in Spain. We had to decide quickly on the conduct of a first test in the sea, which would be the only way to confirm our ideas. It was evident that the

longer one waited to apply biological control, the greater the likelihood   **217**
that the scope of the operation to release slugs, associated with a mapping campaign to determine the locations of *Caulerpa*, would be beyond
the means that we could reasonably count on.

At the beginning of December 1995, we wrote to the Minister of
the Environment calling for the rapid organization of an international
arbitration panel to produce an opinion on the risks associated with introducing the tropical slugs *Elysia* and *Oxynoe* to the Mediterranean.
This arbitration panel could even be conducted according to the recommendations of the international organization based in Copenhagen (The
International Council for the Exploration of the Sea) with respect to all
marine introductions. International authorities would have to be consulted, because the introduction of a new species to the Mediterranean
concerns all the countries that border it. There also had to be an independent assessment, conducted outside the range of possible local or national intervention; the latter could lead to a partial, biased opinion.

The opinion of experts should permit a choice between two op-
tions—either authorize a local test in the sea following a well-devised
strategy, or ask for supplementary aquarium experiments. In the latter
case, we would have had to defer the experiment in the sea until May
1997. If that test turned out positive, an application on a grand scale
could not be arranged before 1998. By then, how far would the invasion
have spread? This question had to be taken into account by the arbitra-
tors, because the loss of time imposed by the desired precautions might
have been enough to negate the efficiency of the solution. The experts
should thus have considered the risks of introduction of the slugs (none,
according to us); were they inferior or superior to the risk of failure
because of loss of time? For a first test in May 1996, we would have
needed authorization before the spring, in order to have had time to
obtain slugs from eggs reared in sterile conditions. We found ourselves
in the same situation as in 1991, when I had presented an urgent request
to the Ministry of the Environment to eradicate an alga that at that
time covered only a few hectares. In February 1996, Manche succeeded
in submitting a file to the European authorities, which was equivalent

218   to the tacit support of his administration. By contrast, no ministerial office wanted to finance the audit, which consisted of airline tickets and expenses for a stay in Nice for two research specialists: an American and a Dane. The assessment of the biological control project foundered on this problem.

It was at this time that the article by the Monaco researchers appeared in the journal of the Academy of Sciences, presenting the invasion of *taxifolia* as a natural event, just in time to justify the prudence of certain decision-makers, who chose to temporize by waiting for the situation to clarify. This was why our proposal for biological control was not subjected to an assessment before the critical date.

In May 1996, more than two years after the arrival of the slugs in our laboratory, without any external support, we pursued our aquarium experiments to understand the biological characteristics of these strange molluscs better. The accumulated results were the best form of encouragement. At the end of May, between Nice and Monaco, the water temperature reached 18°C at the surface; this was the moment to release the slugs. But without the backing of the authorities, we had to forgo the opportunity. A year was thus lost. While waiting for the assessment that would permit us to begin our experiments at the end of spring 1997, we kept our little slugs warm, along with the hopes they represented.

Chiaroscuro: 1997-1998

During these two years, five scientific meetings were held on *Caulerpa taxifolia*. Sometimes equivocal, sometimes clear, the conclusions, opinions, appeals, and recommendations did not cause any rush in France and Italy to take charge of the situation and bring it under control. Academic and administrative obscurantism has smothered the efforts at scientific clarification.

The Colloquium of the Academy of Sciences (Paris, March 1997)

Without taking a position, the editorial board of *Comptes Rendus de l'Académie des Sciences* published, six months apart, two opposing hypotheses on the origin of the alga. We rebutted the first study, conducted by Jaubert and his coauthors from the Scientific Center of Monaco, with a second publication. The comments in the press displeased the academicians, and the resurgence of the public debate on the origin of the alga annoyed the Ministry of the Environment. The extreme confusion was bound to lead somewhere.

During the first third of 1997, the bureaucrats of the Ministry of the Environment hoped to reach a definitive verdict on *Caulerpa taxi-*

folia by holding an international colloquium, the organization of which was assigned to the French Academy of Sciences. To minimize the confrontation, the meeting was intended to be ecumenical. Thus, the opposing parties were invited along with fifty illustrious academicians or foreign colleagues who had been neutral until then. The location of the colloquium was also intended to ease the tension—the prestigious quarters of the Academy of Sciences. The colloquium was entitled: "Invasive species: the case of *Caulerpa taxifolia.*"

To organize the colloquium, the Ministry of the Environment called on the biochemist François Gros, professor and permanent secretary of the Academy of Sciences. The allotted budget was 300,000 francs, allowing the participation of foreign specialists. The colloquium lasted three days. Half the time was devoted to general expositions on the genetic techniques that could determine the origin of *Caulerpa taxifolia* in the Mediterranean. The other themes (impact on biodiversity, toxicity, modeling of population growth and dynamics, cartography, biological control) were each represented by one or two specialists who were each granted 15 minutes to explore a subject.

The conclusions of the colloquium were completely predictable. To know more about the alga, genetic studies were indispensable. In other words, without molecular techniques, our arguments (and those of our opponents) were unfounded. The two adversarial teams had to go away and do their homework. To debate fundamental questions was thenceforth futile.

Despite the strong opposition of several French participants, the data that we had challenged on the metamorphosis of *Caulerpa mexicana* into *Caulerpa taxifolia* were discussed at the end of the colloquium. This discussion aroused as many reactions of indignation as of silence and embarrassment. Jaubert was urged to explain himself. He proposed a new hypothesis in which it was no longer a question of *Caulerpa mexicana* or of metamorphosis, but of two distinct stocks in the Mediterranean—a strain of *Caulerpa taxifolia* with typical fronds and a *Caulerpa taxifolia* whose fronds become "mexicanoid" under the impact of repeated grazing by herbivores. Among the audience were several taxonomists of green algae. I was surprised by their lack of reaction to this new version of transmutation in the genus *Caulerpa*. Everyone seemed convinced that genetic analysis would provide clearer insights.

For our part, we reminded the representatives of the Ministry of the Environment that they had received five requests since 1992 for assistance in genetic research. However, at least in the beginning, the origin of the alga was not in question, as our hypothesis that it been introduced from the aquarium at Monaco was widely accepted.

The director of a laboratory of algal genetics, at Roscoff in Brittany, was both a partner in one of the projects and a referee of the two contradictory studies published by the Academy of Sciences. In 1993, to undertake a genetic study of *Caulerpa taxifolia*, he had requested a million francs, a sum then judged excessive. Without substantial financial support, neither his laboratory nor the numerous other French teams specializing in plant genetics would devote themselves to the study of the alga, whereas, for all the other *Caulerpa taxifolia* projects (toxin chemistry, ecotoxicology, physiology, population dynamic modeling, impacts on biodiversity), the scientific community had rapidly mobilized itself.

During a break in the formal presentations, an hour was devoted to films and slides. Thomas Belsher of IFREMER showed a realistic film entitled, "On the Trail of the Caulerpe." His film has since been made commercially available by IFREMER and produced in three languages (French, English, and Arabic). Michel Denizot, a retired phycologist asked by the Museum of Monaco to evaluate the impact of the alga, presented a film without commentary. It consisted of successive shots of a *Caulerpa taxifolia* prairie photographed every 15 seconds during the same hour every two weeks for a year by the camera installed under the Oceanographic Museum. The alga always covered nearly the entire field of vision. Seasonal morphological fluctuations are evident, and from time to time schools of small fish are visible. Supposedly showing scenes of life above *Caulerpa* prairies, the film primarily revealed the surprising stability and density of the alga. My contribution was to project slides showing the diverse habitats colonized by the alga and to distribute a status report on the invasion.

For his part, Jaubert brought a magnificent little aquarium containing a sprig of *Caulerpa taxifolia* and two rare and beautiful tropical fish. For the three days, this aquarium remained beside the speakers' lectern. The message was clear and unchanged: the alga does not threaten the environment.

IFREMER presented the various means of prevention and management. It was evident that all methods then known were applicable only to small areas. Global eradication was no longer possible. On this point, the academicians and editors of the conclusions of the colloquium thought they had reached consensus for the first time, but the Scientific Committee on *Caulerpa taxifolia* had already reached this conclusion in October 1992, having observed that the invaded area had grown from 30 hectares to 420 hectares in only one year. The conclusions of the colloquium also recalled all the preventive measures that we had enumerated from the beginning of the affair—to alert the public so as to avoid dissemination by humans, to create sanctuaries, etc. This restatement was obviously useful and positive.

By contrast, the response of the colloquium to our suggestion to use biological control was far from satisfactory. Six months before the colloquium, the Ministry of the Environment had responded to the request for support we had submitted a year earlier. Michel Echaubard was assigned to examine our proposal. A specialist in the ecotoxicology of pesticides in terrestrial habitats, a member of the green movement, and well versed in the many resounding failures of biological control in various parts of the world, he produced a very negative report on our plan to use tropical slugs. His main criticisms boiled down to two points: the risks of introducing tropical species into the Mediterranean, and the need for research on the potential use of Mediterranean slugs.

We had already studied these native slugs (*Oxynoe olivacea* and *Lobiger serradilfalci*). We observed that they do not eat much and have pelagic reproductive cycles, which disperse their larvae. These traits work against the establishment of dense populations, which would be needed to control the spread of the alga. It was therefore futile to place our hopes in these species.

At the colloquium, biological control was treated in thirty minutes; **223** fifteen minutes for my exposition, in the course of which I pleaded for research on the two species of tropical slugs that might be useful in the battle against *Caulerpa,* and fifteen minutes for a Danish sacoglossan specialist. She presented the remarkable biodiversity of sacoglossans worldwide and called for preliminary studies of competition between the two species proposed for introduction and the rare Mediterranean natives. Echaubard was assigned to write up this delicate question of biological control in the conclusions of the colloquium. He opposed the introduction of tropical sacoglossans into the Mediterranean.

A press conference was organized at the end of the colloquium. As I had been called a "Zorro of the media" by one of the academician organizers, and not wishing to monopolize the reporters' questions, I declined to participate. Suffice it to say that the colloquium organizers had not appreciated the appearance of the first edition of this book the very day the colloquium began. This was a simple coincidence; the publication date was that of a Paris book fair and had been selected by my publisher well before the colloquium date was set. Of course I had alerted the academicians to the imminent appearance of the book as soon as I learned about the colloquium.

I expressed my satisfaction with the outcome of the colloquium to an AFP reporter whom I ran into as I left the Academy. The conclusions confirmed our views in their entirety. We were correct to have asked for genetic analyses since 1992, correct to have said, beginning in 1992, that it was no longer possible to eradicate the alga, and above all correct in that two-thirds of the colloquium conclusions suggested various means of prevention and management of the invasion. Even though I felt that the threat posed by the alga to Mediterranean habitats had been insufficiently emphasized, *Caulerpa taxifolia* was not judged innocuous or beneficial, as our opponents had claimed. At the press conference, my friends expressed similar opinions to mine. The book and the colloquium were heavily reported in the press. On the whole, the general feeling was that we were right and that we were satisfied with the colloquium conclusions. The academicians and the bureaucrats of the Ministry of the Environment had not controlled the media coverage and had not succeeded in effecting a rapprochement between the opposing teams.

The Colloquium of the International Council for the Exploration of the Sea (La Tremblade, April 1997)

Far from the media and academic debates, a new colloquium on invasive marine species was organized at La Tremblade, 100 kilometers north of Bordeaux in IFREMER quarters, under the aegis of the International Council for the Exploration of the Sea. The theme of one study group was biological control of invasive marine species. Among the few cases of biological control contemplated in the sea and discussed by the study group was our proposed use of the tropical slugs. The meeting, chaired by Jim Carlton (of the Mystic Seaport Museum, Connecticut), a leading international specialist in the problems of marine introduced species, was very constructive. Our research was analyzed and new studies were proposed; on the whole, we were encouraged. One main conclusion was that marine biological control should not be approved or rejected by a single nation. It concerns all the countries bordering a sea or an ocean. When an introduced species invades and damages ecosystems and a biological control species might be introduced to limit its spread, this possible solution should be the object of an international analysis to define the conditions of its possible use. The authorities of the various affected countries should, as a group, adhere to the recommendations of the expert analysts without fear of opposition from other countries.

We regretted the absence of bureaucrats from the Ministry of the Environment, of Echaubard, their designated expert and chief critic of the biological control proposal, and of the co-organizers of the Academy of Sciences colloquium. The Ministry of the Environment had apparently closed the file on the *Caulerpa taxifolia* biological control project. Our tropical slugs, *Elysia subornata* and *Oxynoe azuropunctata*, no longer interested anyone, but that did not prevent us from continuing our research on this possible management tool.

The Third International Workshop on Caulerpa Taxifolia (Marseilles, September 1997)

In September 1997, the third international workshop on *Caulerpa taxifolia* was held at Marseilles by the officials of GIS–Posidonie in charge of coordinating the European program of research on *Caulerpa taxifolia*. About a hundred researchers from throughout the Mediterranean region assembled to listen to reports on research undertaken by several dozen scientific teams thanks to the European "Life" program. Elected officials and representatives of various governmental agencies also attended. Once again, the researchers restated the Barcelona Appeal (see pp. 189–90). But, as at La Tremblade, bureaucrats from the Ministry of the Environment and the academicians who organized the Academy colloquium were absent.

An appeal to Prime Minister Lionel Jospin was drafted by all the assembled participants and was co-signed by Boudouresque and me and by Dolorès Baudelot, representing the mayors of the coastal region of the Department of Var. The sole response was a notification that the appeal had been received, along with a notice that it had been sent to the minister with jurisdiction over the matter, Dominique Voynet, a representative of the green parties and Minister of Territorial Development and the Environment. There was no sequel. During this time, *Caulerpa taxifolia* continued to expand at St.-Cyprien.

APPEAL TO THE PRIME MINISTER

Mr. Prime Minister,

The assembled scientific participants and local elected officials present at the Third International Colloquium on *Caulerpa taxifolia*, held at Marseilles on September 19–20, 1997, call on you to make a firm and clear commitment to battle the spread of *Caulerpa taxifolia* on the French Mediterranean coast.

The alga continues to proliferate and to spread very rapidly, and the means to fight it have been operational for some time. The consequences of the invasion, damaging from both the ecological and socio-economic standpoints, have been clearly established.

The participants thus propose that you quickly establish a global strategy to take charge of the battle to contain the expansion while working closely with the local elected officials and the governments of other affected nations.

An initial action should be to solidify the engagement of the French government in this problem by eradicating a colony of *Caulerpa taxifolia* that has been growing since 1991 in the harbor of St-Cyprien (Languedoc–Roussillon) and that threatens the Spanish coast.

France, the nation currently most affected by the algal invasion, has the obligation to set an example by taking charge of dealing with the problem.

Counterattacks

Once my book appeared, Doumenge and Jaubert filed separate libel suits. Doumenge cited many passages and requested one million francs in damages as well as the publication of the judgment in five newspapers. He lost the case initially but, on appeal, won one symbolic franc as well as publication of the verdict in three newspapers. As for Jaubert, he demanded the immediate withdrawal of eight passages and the publication of the judgment in national and international newspapers at a cost of 160,000 francs. He won one franc in damages and the removal of several passages in future editions of the book.

In December 1997, the popular science magazine *Science et Vie* ran a vitriolic fifteen-page piece contradicting our hypotheses and casting doubt on our motives. The articles were titled, "The ecological and fi-

nancial scandal," and "The profitable *Caulerpa* business." The latter was   **227**
signed by the freelance reporter Jubelin, already cited in this book (on
page 194). The first article explained that the prevailing currents would
have carried the alga from the Red Sea to Monaco. A half-page photograph of the Suez Canal suggested that the Red Sea is connected to the
Mediterranean.[1] According to the author, the alga would have floated
while stuck amidst other mucilaginous algae and thus would have been
carried by the currents; in fact, these algae would sink. The hypothesis
that *Caulerpa taxifolia* and *Caulerpa mexicana* were one and the same
species was reaffirmed, illustrated by a large photograph showing the
great morphological variation of these algae. The subtitle of the article
was, "*Caulerpa* can be innocuous and may even cleanse the marine
bottom."

A large insert questioned a mapping report on several hundred hectares produced by IFREMER and the environmental bureau of Monaco.
The article focused on only a single parcel, monitored by Jaubert, who
organized dives at the site and helicopter monitoring with the aid of a
spectrograph, and by Jubelin, who dove in the company of a certified
bailiff. The article concluded, "The documents seem to show that *Caulerpa* is not so invasive after all."

The second article was still more accusatory; it was subtitled "The
hidden side of a financial scandal" and "The big business of *Caulerpa.*"
The article suggested that the alarmist researchers had concocted the
entire affair, without any scientific assessment, for the sole purpose of
garnering funds for their laboratories. Boudouresque, GIS-Posidonie,
and I were accused of wasting public funds "in a briskly run financial
operation." Our sponsors, the Ministry of the Environment and the European Union, were depicted as incompetents: "The appointed (European) experts are not competent in underseas questions." The allocation
of funds for research carried out since 1992 by more than thirty laboratories, as well as the procedures for submitting files to the European
Union and the Ministry of the Environment, were presented in a confused fashion that led one to believe in a giant scheme: "A textbook case

that shows how to obtain funds—major funds—without having to be accountable for them." The whole was buttressed by citations of anonymous bureaucrats of the Ministry of the Environment. The message was not subtle: "From now on, it's not those who make the most noise who will be proven correct."

The geneticist who had requested a million francs for his laboratory in order to undertake a genetic study of the alga departed from his habitual reserve to say that the entire *Caulerpa* affair was a business operation: "For Bernard Kloareg, the nature of the affair is evident: 'They have exaggerated the impact of Caulerpa for purely mercenary ends.'"[2] Nearly three thousand free copies of the journal were distributed to many oceanographic researchers in France, to diving clubs on the Côte d'Azur, to local political authorities, and to national agencies—paid for by whom?

This matter, which we interpreted as yet another attempt to convince the authorities that the theses defended for seven years by the scientists based in Monaco were correct, did not strike much of a responsive chord in other media outlets. "Information or intoxication?" asked Agence France-Presse in a dispatch entitled, "*Caulerpa taxifolia: Science et Vie* 'dives' into the polemic," which observed that instead of resolving the conflict the editor-in-chief of *Science et Vie* had chosen to release the fox in the henhouse.[3] This was not the opinion of the director of the environment for IFREMER, who stated on a televised news show on France 3: "It is better to disseminate knowledge, well-supported facts, and not to sound the alarm until one knows why, how, what impacts, what are the threats; because science cannot yet really give answers to these matters, it would be better if it refrains from asking the questions."[4] No comment . . .

Of course the articles in *Science et Vie* ended up serving our cause. The alarmist scientists gathered in Marseilles to formulate a common defense strategy. The participants were divided. Some wanted to ask *Science et Vie* for a right of response, others did not want to rejuvenate the polemic. The latter was the course we settled on. We refused to feed the debate, because when two persons argue in public with only the media to hear them and echo them, they end up being equally discredited. Even if we had produced a well-reasoned response, the reader would still be left in doubt; the public does not like to be the captive audience to an angry run-in between scientists. Finally, we feared that our right of response would be accompanied by a commentary and we would not be allowed to reply. Many of our friends wrote to *Science et Vie* to complain about the articles. The only thing to appear was a paragraph from a reader who felt that *Science et Vie* was correct.

Denis Ody, a journalist but above all a former member of Commander Cousteau's team, tried in vain to publish a summary news article. He distributed it over the Internet. Here is his testimony, recently taken up by the press:[5]

At the heart of the article is the problem of the origin of the alga. I should describe my own experience on this matter. When, in January 1992, under orders from Commander Cousteau (indirectly implicated because he was the director of the Museum at the time of these events), I went to the site to understand the problem posed by the growth of *C. taxifolia*, I met a person in charge of the Museum aquaria. This person clearly recounted how, when it was necessary to clean the tanks invaded by *Caulerpa taxifolia*, entire garbage cans full of freshly torn out algae were thrown from the windows into the sea. Why? Simply because the route one would have had to take to carry them to the garbage dump was long and difficult with a heavy, cumbersome garbage can, and no one at that time imagined that a tropical alga could survive in the Mediterranean! It is apparently a current practice, because scientific divers accompanied by the Prince's police officers have noted the presence, always at the foot of the Museum, of tropical coral specimens—in this case, dead ones.

A few months before the publication of the articles in *Science et Vie,* a letter from Cousteau, dated 1994, appeared in the press. He had sent an unequivocal plea to the Minister of the Environment, Michel Barnier, which closed, "Its proliferation on a large scale would lead to a

230   notable diminution of biodiversity and of the richness of coastal ecosystems as well as to a radical transformation of the underwater landscapes
that were the scene of my first dives."[6]

The Workshop of the United Nations Environmental Program
(Candia [Heraklion], Crete, March 1998)

In March 1998, the problem was raised to an international level by the
United Nations under the umbrella of its environmental program, and
more particularly its Action Plan for the Mediterranean. This was a
high-quality meeting: nineteen of the twenty-one countries bordering
the Mediterranean were represented by scientists officially appointed
by their governments, with only the Serbs and the Bosnians absent.
Monaco sent Jaubert. France was represented by both a member
of IFREMER (Thomas Belsher) and a bureaucrat from the Ministry of
the Environment (Vincent Bentata). The main scientists involved in the
previous programs on *Caulerpa taxifolia* had been invited. Jubelin was
the only registered reporter, sent by *Science et Vie*.

The colloquium, with a very diplomatic title ("Workshop on the
Caulerpa species invasive in the Mediterranean"), encompassed the
study of problems posed by all Mediterranean *Caulerpa* species. All
the nations bordering the Mediterranean were thus concerned, for there
were *Caulerpa* species off the shores of all of them. Moreover, another
species of *Caulerpa* (*C. racemosa*) had begun to manifest an invasive
tendency, especially in the eastern Mediterranean, over the last ten
years. At the end of the discussions, the workshop participants agreed
that *C. taxifolia* and *C. racemosa* posed a grave threat: they might upset
an ecological equilibrium. Several recommendations were unanimously
adopted. All the conclusions and recommendations of the workshop
were adopted by *all* the participants.

Influenced by the Convention on Biological Diversity adopted at
Rio de Janeiro and the 1995 Barcelona Convention, the workshop recommended that all nations take "all appropriate measures to regulate

the intentional or accidental introduction of non-indigenous or genetically modified species . . . and prohibit those that may have harmful impacts on the ecosystems or species" and endeavor "to implement all possible measures to eradicate species that have already been introduced
when, after scientific assessment, it appears that such species cause or
are likely to cause damage to ecosystems, habitats, or species in the area
to which this Protocol applies." With respect to *Caulerpa*, they urged
all Mediterranean countries to cooperate to stem the spread of *Caulerpa taxifolia* and *Caulerpa racemosa* in the Mediterranean and to support research and information exchange on these two species. Most
strikingly, the workshop called for the prohibition of the sale and use of
Caulerpa taxifolia and *Caulerpa racemosa*, and a halt in the use of all
Caulerpa in aquaria (except for the Mediterranean species *Caulerpa
prolifera*). Finally, the participants recommended the public distribution of information on how to avoid spreading *Caulerpa*, the eradication
of small colonies from ecologically important areas, and research, mapping, and monitoring to understand the trajectory and threat of the invasion.

The Meeting of the Marine Conservation Biology Institute
(Leavenworth, Cascade Mountains, Washington, USA, May 1998)

A non-governmental organization, the Marine Conservation Biology Institute, which includes many American scientists, organized a meeting
on the theme of biological control against marine pests; I was invited.
Once again the case of *Caulerpa taxifolia* and the possible use of tropical
slugs was explored. Not only were the two leading specialists on sacoglossans invited (Kerry Clark of the Florida Institute of Technology and
Kathe Jensen of the Zoological Museum of Copenhagen, Denmark), but
the organizers were able to assemble phycologists such as Richard Norris
of the Smithsonian Institution and many specialists in biological control
of aquatic and terrestrial pests in the United States and Australia. In
highly productive debates, the dangers and opportunities of biological

232 control of *Caulerpa taxifolia* were examined. I was listened to more closely, better understood, and better supported than in my own country, and I was encouraged in my initiatives by these major specialists.

Struck by the story of the invasion of the Mediterranean by *Caulerpa taxifolia* and aware of the potential dangers posed by this remarkable strain that is now commercially available worldwide, a group of participants at this workshop drafted a letter to the United States government calling on it to prohibit the use of *Caulerpa taxifolia* in American aquaria (a measure that had just been taken by Australia). Signed by 107 researchers and specialists, the letter was sent on October 21, 1998 to U.S. Secretary of the Interior Bruce Babbitt and noted in the October 30 issue of the journal *Science*. Acting remarkably quickly, the U.S. Department of Agriculture, on December 4, 1998, proposed adding the Mediterranean clone of *Caulerpa taxifolia* to the Federal noxious weed list, an action which would prevent the importation and interstate movement of this alga, but not intrastate movement. The U.S.D.A. called for comments on their proposal through February 2, 1999; on April 15 of the same year, the clone was listed as a noxious weed.[7]

The Action Plan of the French Ministry of the Environment

Despite the recommendations of the Academy of Sciences and the United Nations Environmental Program, the French and Italian governments made no attempt to organize an approach to prevention and control in 1997 and 1998. Spanish and Croatian authorities continued to support operations to control the algal spread. For the first time, an eradication was attempted in Monaco in 1997 on a cliff of a red coral reserve.

In France, two members of the Chamber of Deputies led the charge and posed written questions to the Ministries of the Sea and the Environment (in the government of Lionel Jospin, the latter had become the Ministry of Territorial Development and the Environment). These queries related to the slow development of preventive measures. The director of the Port-Cros National Park raised this same issue with his ministry (Environment). These three initiatives received similar ambiguous responses: the alga in the Mediterranean arose from an introduction or spontaneous "appearance"; it was necessary to know more about the origin before acting.

During the summer of 1998, the Ministry of the Environment created a new scientific committee. No researcher who had previously studied *Caulerpa taxifolia* was included. The committee was charged with analyzing the responses to a request for proposals for new research on *Caulerpa taxifolia*. Without any interaction with scientists who had worked on the subject for more than seven years, the committee formulated a new program of studies on *Caulerpa*. The deliberations of this committee were not informed by the advances of the program already underway, financed jointly by the European Union, the Provence–Alpes–Côte d'Azur region, and the Ministry of the Environment, because the many results of these studies had not yet been reported. The government plan was to start at ground zero with new personnel, and conduct new research before undertaking the control and management measures advocated by the Academy of Sciences, even if it meant financing two overlapping research programs. Vincent Bentata, in charge of the *Caulerpa* affair for the Ministry of the Environment, first announced this new policy in the piece in *Science et Vie* discussed above. He told the reporters that his ministry was going to initiate an action plan with a budget of four to five million francs composed of three parts: new research on the impact and genetics of the algal invader; a precise cartographic analysis; and measures to exclude and to fight the alga, bearing in mind that "the means of fighting it obviously depend on the results of the first two parts of the action plan." This whole policy seemed like a surreal flight to the future; new research was initiated that would surely lead to a new colloquium that would suggest new research orientations. At this time, about 300 publications or scientific reports had already appeared on *Caulerpa taxifolia*, which thus had become the most documented example of an introduced marine species.

The marginalization of the researchers who had invested so much in this affair seemed to be another undeclared objective of this plan,

234 but one that was announced by our most determined adversaries. I was directly affected. In June 1998, I learned that the two university laboratories at Nice (of which mine was one) that had published on *Caulerpa taxifolia* were not recognized by the Ministry of Research and Higher Education, entraining for at least two years a decrease in space, funding, promotions, and recruitment.

Scientific Clarifications

In the autumn of 1997, despite the recommendations of the Academy of Sciences, no genetic research had been initiated. In the laboratory we still maintained a collection of *Caulerpa* gathered in diverse, distant sites in the Mediterranean (Balearic Islands, Croatia, Sicily, etc.) and tropical seas (the Caribbean, southern Japan). Samples of *Caulerpa* were sent to three genetics teams that wished to helped us: one to the United States, to a young *Caulerpa* researcher (Rich Zechmann, then at Duke University), the second to Japan (to Teruhisa Komatsu, who interacted with a genetics laboratory), and the third to Marseilles (to a team affiliated with Boudouresque). But none of the three teams had had the time to analyze our samples. A brilliant student—Olivier Jousson—allowed me to make progress on this front. A doctoral student at Marseilles, working on the genetics of marine parasites, he was curious and tested a fragment of *Caulerpa* that he had collected in the sea. He easily obtained, at little cost, a very nice gene sequence that is presently the most useful for distinguishing populations and species of *Caulerpa* (the gene known to geneticists as ITS rDNA). Bolstered by his success, he got the support of his original laboratory, in Geneva, to examine my entire collection. From Geneva, he easily got other samples from aquaria in eastern France (Nancy) and Germany (Stuttgart) from which *Caulerpa* had been imported to Monaco in 1982. In November 1997, all twenty-four samples were prepared: seventeen *Caulerpa taxifolia*—ten from the Mediterranean, five from aquaria, and two from tropical seas; three *Caulerpa mexicana*, including one from the Israeli Mediterranean coast;

and four from other *Caulerpa* species. The results were in just a month   **235**
later. They were so astonishing that the analyses were repeated. The
evidence was compelling: all the *Caulerpa* samples from the aquaria
and from the Mediterranean were the same. Moreover, the *Caulerpa*
taxifolia all differed greatly from the *Caulerpa mexicana*. These findings
confirmed our hypotheses based on direct observation, the dynamics
and the spread of the invasion, and the completely characteristic traits
of the strain of *Caulerpa taxifolia* in the aquaria and in the Mediterranean. But what surprised everybody was that no genetic variation whatever was detected. In the absence of all sexual reproduction, the study
suggested that the aquarium strains and those from the Mediterranean
were just a single, extremely vigorous and resistant individual, fragmented and cloned to infinity.

   This remarkable result was written up in a scientific article in early
spring 1998. The article was submitted for publication to the scientific
journal *Nature* twice. It was rejected. Three months had been lost. It
was then sent to one of the most prestigious scientific journals in marine
ecology (*Marine Ecology Progress Series*), which quickly accepted it. It
was published in October 1998 and reported widely by the press.[8] The
geographic origin of the alga coming from northern Europe in the 1970s
remains to be determined. A major inquest has been initiated on this
point and it is closing in on the answer.

   In the meantime, in early 1998, we learned that a team of French
geneticists from Lille had been asked by the Ministry of the Environment to study the genetics of the alga. Research funds and a scholarship
were placed at their disposal in late spring 1998. This decision was
reached without any interaction with our laboratory, which had so frequently asked for assistance for the genetic analysis of our collection.
We learned moreover that the Lille researchers were not supposed to
collaborate with us! They thus did not use my collection and began taking samples on the Côte d'Azur in June 1998. We felt marginalized,
nearly shunned. But I concede that a verification of the first study on
the genetics of *Caulerpa taxifolia*, by a neutral team that had collected
its own samples, could be useful. Moreover, the first publication of this

team, collaborating with a Dutch genetic laboratory at Groningen led by a specialist in the genetics of green algae (Jeannine Olsen), appeared at the same time as ours.[9] The research was conducted on four samples of *Caulerpa mexicana* and five samples of *Caulerpa taxifolia*—one from Australia and four from the Mediterranean, of which three were provided by Chisholm and Giaccone, who adhered to the hypotheses of the scientists based in Monaco and who labeled them ambiguously as "*C. taxifolia–C. mexicana*." The genetic analysis was done using the same technique as ours (sequencing of the gene ITS rDNA). The conclusions too are the same: *Caulerpa taxifolia* and *Caulerpa mexicana* are not the same species, and the *Caulerpa* labeled as "*C. taxifolia–C. mexicana*" are only *Caulerpa taxifolia* identical with others from the Mediterranean (including one from Sicily). On this last point the two genetic studies confirm our taxonomic analysis based on the simple morphological features used since 1849 by the great majority of phycologists.[10] This second publication on the genetics of *Caulerpa* did not include samples from aquaria and thus remains ambiguous about the origin of the alga in the Mediterranean.

The appearance of the first publication on the genetics of the alga marked the end of all ambiguity about the origin. The alga henceforth has the unquestionable status of an introduced species and is treated as a pest. This news also unleashed a new political initiative. An influential scientific reporter (Françoise Simpère) has called on the Minister of the Environment and the parliamentary deputies who are divers to enact immediate measures of prevention and control. The deputies reacted quickly, and, remarkably for the Fifth Republic, deputies from all parts of the political spectrum worked together to bring forward a law calling for all necessary measures to battle this invader quickly. Divers, fishermen, and conservation organizations demanded a parliamentary inquiry on the basis of the polemic and the slow administrative response over the past eight years. In late December 1998, the Chamber of Deputies committed verbally to undertake such an inquiry, and, in January 1999, a motion to that effect was formally introduced. Scientific knowledge has finally informed political action!

During this period, the alga has continued to spread. Its impact, so obvious for all those who had seen it *in situ*, is still greatly underestimated by the authorities in the affected countries. For skeptics, it is worth recalling that leading ecologists consider invasive introduced species to be a growing threat to global biodiversity. They believe that habitat destruction and invasive introduced species are the two greatest threats to biodiversity,[11] a conclusion drawn at a 1996 United Nations conference on introduced species. In April 1997, more than 500 American scientists called for the creation of a presidential commission to recommend strategies to prevent the introduction of invasive species and to control them once they are established. E. O. Wilson, the biologist who has most surely marked the end of this century in the area of biodiversity protection,[12] observes that the extinction of species by habitat destruction can be compared to a violent death in an automobile accident, while the extinction caused by invasive exotic species is a slow dying by disease: it is progressive and insidious.[13]

In July 1998, we published a voluminous report on the spread of the alga throughout the Mediterranean, the fruit of thousands of dives by thirty-five researchers aided by hundreds of volunteers.[14] An observation network now exists, resulting from an exceptional collaboration between the scientists and divers, yachtsmen, and fishermen. For the first time the invasion of a marine exotic species has been fastidiously mapped year after year since its introduction. By the end of 1997, ninety-nine independent algal invasion sites were inventoried (as opposed to sixty-eight a year previously), the area dominated by the invasion (more or less occupied by *Caulerpa* colonies) reached 4,630 hectares (it had been 3,052 hectares at the end of 1996), and the linear extent of coastline occupied by *Caulerpa* reached 83 kilometers (it had been 58 kilometers at the end of 1996). During the summer and autumn of 1998, dozens of new sites or expansions of existing ones were reported to my laboratory. The arrival of *Caulerpa taxifolia* 1.5 kilometers from the heart of the Port-Cros National Park, thirty-nine meters deep near the magnificent island of Gabinière, aroused great anxiety among all divers on the Côte d'Azur.

238   In August 1998, Thomas Belsher of IFREMER was again assigned to map the spread of the alga in its introduced range. Dozens of kilometers of the sea bottom were filmed by an underwater camera hauled by the oceanographic vessel *Europe*. The alga continued to grow. In October 1998, for the first time, Belsher organized a joint mission between France and Italy. On board the *Europe*, Italian researchers participated in the cartography of the algal invasion along the Italian coast of Liguria (between Genoa and Monaco). The observations were not surprising—new colonies discovered, important invasions detected, etc.

It is only a matter of time before the alga spreads through the entire Mediterranean. *Caulerpa taxifolia* is indeed a stranger that has colonized paradise. At the beginning of the 1960s, Rachel Carson dedicated her book *Silent Spring*,[15] a ringing plea against pesticide pollution in the United States, to Albert Schweitzer, who had written: "Man has lost the capacity to foresee and to forestall. He will end by destroying the earth." This epigraph is just as appropriate today.

CHAPTER 7

The Three Lessons of *Caulerpa*

"Announce, denounce, and never renounce"
("Énoncer, dénoncer, ne jamais renoncer")

RENÉ RICHARD[1]

Throughout this book, the reader has witnessed the failure of my many attempts to publicize a major threat to the environment. By its remarkable adaptation to the Mediterranean, the alga has alas proven me correct. But this work is not a simple autobiographical essay of a Cassandra satisfied to see his prophecy confirmed. It is rather the exposé of a double defeat—on the one hand, the loss of control of the alga, which spreads each year and drastically modifies marine life, and on the other hand, the debacle of scientific reasoning, which failed to carry the day in the face of flagrant deception by certain scientists and the laxity of the powers in office at all levels.

Out of bitterness or rancor, I could have been content to denounce the carelessness or culpable negligence of humans towards nature. But that would not have taken me very far; abhorrent behaviors are inherent in human nature; they manifest themselves repeatedly and have often been described.[2] These past years, I have tried to understand the reason for so many misunderstandings, fruitless controversies, and vain polem-

240 ics. I have become convinced that the explanation must be sought elsewhere than in the deplorable motivation of several individuals. It all adds up in my mind to one question: How was the fallacious thinking of sophists able to fool so many people for so long and with so little effort?

This question leads back to three reflections or observations. The first concerns the communication of scientific research results. The credence granted by decision-makers to casual discussion by influential persons in media settings equals if not surpasses the attention they pay to scientific publications. In the absence of ethics and critical scientific reviews, the door is open to the worst sorts of exploitation of new modes of disseminating knowledge. However, in the *Caulerpa* affair, appeals to the media were sometimes indispensable.

The second observation is that of the decline of certain relevant biological disciplines, including ecology, botany, and zoology—the sciences treating nature. For two decades, the study of biology has focused on the chemistry of life; it has become reductionist. In doing so, it has lost contact with its roots; many biologists have become highly competent in molecular biology but have failed to grasp the role species play in nature. The triumph of biological reductionism has been heralded at the very moment when a new notion has been shown to be essential— the diversity of life or "biodiversity." How can biodiversity be managed? How can it be protected when entire areas of knowledge have disappeared? This is my third reflection. This distressing situation is largely responsible for hesitation and complacency in favor of ideas that suggest the easiest and most reassuring responses. For if it is true that critical intelligence is not tied to the quantity of accumulated knowledge, it is also certain that scanty knowledge weakens analytic capability and favors deception.[3]

The tumultuous history of a small alga has illustrated three themes for this conclusion: the diversity of life and its protection, the decline of the relevant natural sciences, and finally the trajectory and stakes of scientific communication. To misunderstand these three principal components can lead to many other conflicts and situations that future generations will regret.

To Preserve Biodiversity

This story of algae and humans will seem pathetic or even shocking to some readers. So much passion, so many discussions, such a lot of work, so much spent in vain over a small alga that, moreover, lives underwater, out of sight for most of us, and, at first blush, seems not to affect the economy! How and why has it so dominated the media at a time when wars, cataclysms, political battles, and economic and social crises make daily headlines?[4]

An alga that disturbs the natural marine equilibria leads to a fundamental concern—the preservation of the diversity of life. Global concerns arose with the threat of extermination of symbolic animal species, such as the panda, baby seal, or dolphin. Today, concern has extended to the ensemble of living species, animal and plant, that populate the earth. The destruction of distant equatorial forests, causing, among other things, the extinction of myriad insect species destined to remain unknown, cries out to us. From the microscopic unicellular organism to the rhinoceros, from the little creek, meeting place of migratory birds, to the Aral Sea, which threatens to dry up, humans are increasingly conscious of their responsibility to maintain biodiversity for future generations.[5]

This awareness has grown with the frequent observation of assaults on the environment and the emergence of problems affecting the entire planet. The diverse forms of degradation combine to modify natural habitats imperceptibly. But it is often difficult to recognize real forms of environmental damage. Like it or not, we get used to the polluted air of our cities or to the water polluted by fertilizers and pesticides spread to excess over agricultural land. By contrast, the destruction of a landscape and the disappearance of a species are strong, brutal warning signals. Never, since the first appearance of life on earth, have so many species disappeared in such a short time; never has the expansion of a single species—humans—changed so many ecosystems in one century.

If the current biodiversity situation is worrisome, the management of biodiversity will be even more difficult in decades to come. Demographers predict a doubling of the global human population over the next century.[6] The first quarter of a century of the third millennium will determine the fate and condition of humanity. In the face of the population explosion that will inexorably exacerbate environmental pressures, safeguarding our natural heritage has become a growing obsession. There is still time to act before we are reduced to destroying everything just to survive. On all sides, international institutions preach "sustainable development" of the planet, which means development for humanity without foreclosing on the vital environmental needs of future generations. The question is both ecological and political, as it calls into question our current modes of development, which are responsible for degrading the biosphere and exhausting natural resources. By pronouncing themselves in favor of sustainable development, the one hundred fifty heads of state represented at the Earth Summit, held under the aegis of the United Nations at Rio de Janeiro in June 1992, have crossed an important threshold in the recognition of the dangers that threaten the planet and its inhabitants.[7] The Rio conference sanctioned the engagement of scientists who had been alerting the public for years to the degradation of nature caused by human activities.

The introduction and spread of *Caulerpa taxifolia* in the Mediterranean has shown once again that the process of degradation, in this case of coastal ecosystems, can be quick. This biotic pollution is disquieting because it is irreversible and is amplified from year to year. Moreover, it strikes at a precious patrimony—the marine life of the Mediterranean, one element of the biodiversity of our planet. It is the protection of marine life that drives me to publicize the risks associated with the invasion by this beautiful but ever so hostile alga.

BIODIVERSITY, A SCIENTIFIC AND POLITICAL STAKE

Originating ten years ago, the concept of biodiversity has been at the heart of concern over environmental management since the Rio conference. It is the study of habitats—ecology—that has allowed the measurement of the harmful effects of human activities and has led us to recognize the need to preserve biodiversity.[8] Because the knowledge of life on earth has numerous gaps, the "precautionary principle"[9] was adopted at Rio: "When there are threats of serious or irreversible damage, lack of full scientific certainty shall not be used as a reason for postponing cost-effective measures to prevent environmental degradation."

The concept of biodiversity has reigned since the end of 1992. It has borne fruit in the publishing world as in research programs imposed by scientific directorates. The term—scientifically correct—succeeds the philosophical notions of the "natural contract,"[10] "Gaia,"[11] and the "new ecological order."[12] It has become the rallying cry of all those concerned with the environment, as widely used by research organizations as by ideological ecology movements, economic powers, or philosophers of the environment. The concept is pragmatic and technical. It allows specialists in ecology, botany, zoology, or microbiology to affirm their role in the hackneyed and too often irrational debate over the protection of nature.

Science against the irrational—that is also what was trumpeted in a petition, paradoxically very hostile to ecology, known as the "Heidelberg Appeal." The day before the signing of the convention to safeguard biodiversity, a group of Nobel laureates, joined by a cohort of less well-known scientists (honored to appear beside their illustrious colleagues) and supported by various multinational industrial firms, sent an appeal to the heads of state and their representatives assembled at Rio de Janeiro. They expressed concern over "the emergence of an irrational ideology which is opposed to scientific and industrial progress" and warned "the authorities in charge of our planet's destiny against decisions which are supported by pseudoscientific arguments or false and nonrelevant data." The message was clear: have confidence in science and industry to resolve the problems of the planet. At the very moment when a more rational management of the planet was going to be established, this appeal sowed discord among the ecologists assembled at Rio to discuss the environment and development. Certainly, these scientific ecologists[13] had always firmly rejected ecological irrationality and could only adhere

244   to some of the intentions formulated in the appeal, such as the defense
of "a scientific ecology centered on taking into account, controlling, and
preserving natural resources." But they strongly condemned this new
form of scientific fundamentalism, noting that the areas of expertise
and the concerns of most of the signatories of the appeal were very far
from the preservation of nature.[14] Who authorized these new inquisitors
to deem what are correct scientific criteria? This is particularly important because, on questions touching on major environmental problems,
the role of scientists is not to render decisions but to elucidate the problems for policy-makers and citizens. Still, the Heidelberg Appeal accomplished the impossible—it rallied the community of scientific naturalists (of whom no member is a Nobel laureate, because their science
is not one of the disciplines glorified by this prestigious prize). This
community unanimously denounced the Heidelberg Appeal as a ham-
handed maneuver of latter day inquisitors.[15]

A BALANCE SHEET OF THE LIVING

The preservation of biological diversity requires first a quantitative assessment of our natural heritage. In this balance sheet, the biodiversity
counted as original assets corresponds to what humans inherited; the
species we have squandered are counted as expenditures. The difference
is what remains, a fragile equilibrium of increasingly threatened lives.
We have to inventory our natural heritage and decipher the interactions
between species or groups of species.

The diversity of life should be studied and understood at all levels.
The first is that of genetic diversity of species. Humans, the apple, the
red coral of the Mediterranean, the bacterium *Escherichia coli*, etc., are
all very different species. Genes transmit their distinctive characteristics
from generation to generation. And within each species, individuals are
not all identical, again thanks to genetic diversity. The varieties of color,
form, and taste of apples correspond to as many different combinations
of genes within the same species. It is thanks to this within-species genetic diversity that crosses between individuals do not automatically

lead to the degeneration of the species. It is also the genetic diversity of a species that allows its evolution:[16] chance matings lead to the emergence of pronounced differences. These are the reasons the massive cultivation of just a few selected varieties of a species risks leading to its genetic deterioration. The abandoned varieties should be preserved, as they constitute a capital of genes indispensable for the production of the species of tomorrow.

The second level is that of the diversity of species. The species is the basic unit of biodiversity. Each species is represented on this planet by populations of similar individuals, capable of mating with one another and of producing fertile offspring. Among the flowering plants, the orchids have the greatest number of different species. The approximately 20,000 species known worldwide in the Orchidaceae all have the same floral architecture but they never cross—they are different species.

The diversity of ecosystems is the third level. The physical habitats and the species that live in them comprise fragile entities in equilibrium with the conditions of life. These assemblages can change as dictated by the evolution or population dynamics of their component species. The variety of forest types—from rainforest to boreal—corresponds to as many distinct ecosystems.

After the inauguration of the biodiversity craze at Rio de Janeiro, other rubrics have been proposed. To the genetic diversity level has been added that of the ensemble of all molecules of living organisms. The ecosystem biodiversity level has been completed by the addition of the diversity of landscapes, indeed even of human communities adapted to a particular set of habitats (the Eskimos or Amazonian Indians). Biochemists, geographers, architects, and anthropologists have thus attempted to incorporate their disciplines in the tally.

In the great accounting book of nature, each living element corresponds to a line. Each gene, species, or ecosystem is unique, thus priceless because its destruction would be irreversible. It is important not to concoct fanciful solutions; no futuristic technology will allow us to reconstruct what has been lost, to cause an extinct species to be reborn. *Jurassic Park* and *Frankenstein* belong to the realm of science fiction.

Protecting biodiversity requires above all a very intense effort to improve existing knowledge in the natural sciences. The extent of gaps in the inventory of species is astounding. Ignorance is enormous and the task immense.[17] It is not a question of several missing lines or several omissions attributable to an accounting error, but of a gulf, indeed an ocean of ignorance. Let us summarize the results of the audit.[18] Under the heading "species," some 1,413,000 living organisms have been inventoried, of which 1,032,000 are animals and 248,000 are plants. But the number of species in fact is much greater: between ten and thirty million organisms. One expert has even advanced the figure of one-hundred million. If we believe the scientists, we know only 1 to 10 percent of the species living on this planet, and we don't even know how much we don't know. While the computers of the United States can verify the identity of one individual among 230 million inhabitants in several seconds, in an era when the Mormons are compiling a register of the genealogies of hundreds of millions of living and dead individuals, at a time when the sequencing of some 100,000 genes of our human heritage is under way, how can we admit that humanity is not yet able to identify all the species with which we share our planet?

For terrestrial habitats, it is especially among the insects and microorganisms (bacteria, microflora, and microfauna) that several million species are still undiscovered. In the sea, marine bacteria and plankton less than two microns long ("picoplankton") are also chasms of ignorance. But all groups warrant consideration; even today we still discover new species of trees, whales, and primates. The inventory should also take account of the status of populations of each species, of its geographic range, of the number of individuals, and finally, by genetic analysis, of prospects for its future.

As for ecosystems, their mechanisms of functioning are yet to be fully understood. Aside from the impacts of physical environmental conditions or human interventions, the diversity of an ecosystem depends on the number and genetic diversity of its component species, of the interactions among these species, and the interactions between each community of species and the neighboring communities. In certain ecosystems, some species are key; all others depend on them. These are most often the dominant plants, such as the oak species in an oak woodland or the pine species in a pine forest. But a tiny insect, the only one able to pollinate the most common plants of an ecosystem, can also be an essential keystone for the equilibrium of the whole. The survival of ecosystems may also rest on the biodiversity of its components; if this biodiversity falls below some threshold, the ensemble may collapse. We are far from understanding the most representative ecosystems of the planet; nor do we know enough about their weak points in order to protect them better.

These observations on our ignorance are all the more unnerving because a single dominant species—humans—is destroying entire swaths of natural habitats, thus causing the extinction of innumerable species before they have even been inventoried. Here it is perhaps a species of little fly that has been definitively eliminated, there an "insignificant" mushroom. Species that are negligible today, but who can say if tomorrow they might have proven indispensable? Would they have had molecules of pharmaceutical interest? Do they play useful roles in their ecosystems? Are they an effective defense against some threat? Who can say? This concern for species that appear insignificant does not arise from the philosophy of the "deep ecologists"—those committed ecologists who ascribe a consciousness or even a soul to all living things—nor from any romantic sensitivity or anthropomorphic compassion, nor again from any partisan interests defended by opportunistic environmentalists. The question is precise, forceful, scientific, and pragmatic.

We have to complete the balance sheet. Clearly, we have to inventory the living and come to an understanding of the life of an ecosystem and how it adapts to change within itself and in its relationships with neighboring ecosystems. We have to learn the weak points of the whole. This is a vast research program that was begun in the eighteenth century, then became obsolete during the last decades before reviving as concern grew for the fate of the earth.[19] By chance, this research has been undertaken again at a time when means of compilation (computers) and communication (Internet and the Web) allow us to rise to the challenge. The only thing now missing is people competent to collect the data. Now, botanists, zoologists, and ecologists are, in many countries, also endangered species. The accountants of life, the identifiers of species (called systematists or taxonomists), are currently considered in scientific circles as old-fashioned biologists, as we shall see. This tendency, which has caused much damage, has barely begun to be checked in the English-speaking world, but not in France. We will soon see that the Trojan horse (the concept of biodiversity) is in place, but that it contains only a few poorly armed warriors.

Protection Measures

Biodiversity may be fashionable today, but environmental protection is not a recent preoccupation. Legislation exists at the national as well as the international level. Lists of rare species have been drawn up and legal measures enacted to protect them. Areas naturally rich in flora and fauna are preserved. In France, many levels of protection and management of protected areas have been defined: national park, regional park, natural reserve, biotope reserve, ZNIEFF (natural zone of ecological, floral, or faunal interest), not to mention greenspaces defined in community land-use plans (POS). At the international level, the most beautiful ecological assemblages have been labeled "biosphere reserves" under the aegis of UNESCO. In the future, a better knowledge of ecosystem function should permit better prediction and correction of human impacts by virtue of protection or management aimed at several key species.

Presently, the efforts are aimed primarily at inventorying habitats and species that seem the most threatened. In Europe, we are inventorying national sites harboring habitats and/or species of community interest,[20] with the goal of assuring a coherent network for the conservation of nature and biological diversity under the auspices of the European Union. An international institution—the International

Union for the Conservation of Nature (IUCN)—has taken on the responsibility of classifying threatened species.[21] The status of threatened
species is granted to certain species that appear headed for extinction;
efforts are then made to protect them.

Such species are classified in several categories. The first are extinct
species. To translate the English word "extinct," the French have preferred the euphemism "éteinte" (which literally means "put out," as a
light is put out, connoting that it might be relit) to "disparue" (disappeared, connoting irreversibility), although the hope of ever seeing a
species again once it is classified as éteinte is very dim. Every species
that was formerly known but has not been encountered in nature for
fifty years is put in this category. Next come endangered species (those
on the verge of extinction). They have been exterminated in great numbers or their habitats have been damaged to a critical level. Action is
required to save them. The term "vulnerable" is given to species of
which most populations have been destroyed by humans and that risk
being classified in the next highest category (endangered) if the unfavorable conditions persist. Finally, rare species are those for which the
population sizes are greatly reduced throughout their ranges for natural
reasons. Specialists have even planned two catch-all categories: the "in-
determinate" category, which groups species belonging to one of the
three higher categories (endangered, vulnerable, rare) but for which we
cannot state the level of threat, and "insufficiently known," which des-
ignates species that "appear" threatened but for which there is simply
insufficient information.

Based on these definitions, international, national, and regional
laws have been enacted with the goal of increasingly better protection
of threatened species. The first to have been protected were among the
most visible, the most beautiful (the elephant, the giraffe, the panda,
the mountain lily); more mundane species have subsequently also been
protected (a beetle, a Mediterranean euphorb). But the extent of these
efforts is closely tied to the number of competent researchers who can
identify threatened species, and the distribution of such people among
different groups of organisms is heterogeneous—the number of experts

250   is very high for species the size of a sheep, but reduced sometimes to one or two people in the entire world for certain groups, like some worms. All species are not equal before the laws of protection.

TRUE AND FALSE THREATS TO MARINE LIFE

The delay in development of knowledge is even greater for the marine realm. The habitat has barely been penetrated, the numbers are poorly assessed, and the habits of species not cultivated by humans are barely known. The relationship between contemporary humans and marine life is at the stage of that between paleolithic humans and terrestrial life. Tribes of hunters (present-day fishermen) track wild prey (now fish and marine mammals) with increasingly efficient arms (sonar, drift-nets). The overexploitation of natural resources is leading humans to cultivate (algae in southwest Asia) or rear (molluscs, crustaceans, and fishes) marine species.

But if the cod or herring stocks fall, these species are not threatened with extinction: it is simply that their catch is no longer profitable, throwing many fishermen out of work. Only a few marine species are known to be threatened with extinction, for the most part "flagship" species (dolphins, whales, seals, marine turtles). For the whole of the Mediterranean, fewer than twenty poorly known species are classified as threatened and are thus protected in some countries.[22] Moreover, marine ecosystems particularly rich in species diversity are better and better protected—among these are coral reefs, estuaries, and seagrass meadows (notably *Posidonia, Cymodocea,* and species in the Zosteraceae).

When changes in the marine environment are discussed, we should pose two questions. What are the principal causes? How important and widespread are the impacts? The first cause to be condemned is pollution. In fact, we are interested in pollution especially when it can directly or indirectly affect humans. We all know that our health can be affected by simple contact with sea water and sand on beaches because of pathogenic bacteria, fungi, or viruses that may grow there. We dread the difficulties associated with ingesting toxic seafood. Another form of pollution denounced by society is the black tides along our coasts, which have quantifiable economic consequences for fishing and tourism.

For all these forms of pollution that threaten our well-being or our economic activities, we have established standards and laws. They define the concentrations of toxic substances or bacteria considered harmful. They aim to limit all deliberate pouring of toxic substances into the sea. Monitoring networks for certain pollutants and bacteria have been established in many regions; these require costly campaigns that mobilize laboratories to conduct detailed analyses. At the national level, campaigns to measure certain pollutants (heavy metals, pesticides, detergents, hydrocarbon derivatives, nitrates, phosphates) are conducted in the sea. But the analysis and interpretation of millions of data, for pollutants that are usually present in unquantifiable traces, are hardly satisfying. We sometimes prefer biological concentrators, for example, mussels (which continuously filter the water) or fishes (located at the tops of food chains), in which one can assess the concentration of substances toxic (to humans) or of molecules indicating the symptoms of physiological stress (the "signature" of the presence of a pollutant). The economic repercussions of bacterial pollution have inspired coastal communities to construct increasingly efficient water purification stations. These moves are aimed above all at guaranteeing highly sanitary water quality for tourists using the beaches; the concern is not protecting marine life.

Pollution harmful to humans should not be confused with impacts on the diversity of life. We do not know the thresholds of tolerance of species and habitats to many pollutants, and we still know little about their effect on the complex, structured life of the sea. By contrast, other changes for marine life, better evaluated, are much more harmful than the pollutants most feared by humans. The habitat conversion of coastal zones and some fishing practices—activities that are lucrative for humans and not considered as pollution—are among the worst forces affecting coastal marine species. To these degradations one should add biological pollution, engendered by introduced species that invade and disrupt certain ecosystems. Human activities that destroy marine life are usually neglected and are not the target of protective measures. We

252   content ourselves with estimating the damage and describing the various types of impact, while the causes (chemical, physical, biological, or interactions of different sorts) are not understood in detail. The assessment of marine life faced with pollution is summarized in several "health bulletins" for certain easily observed and quantified species or ecosystems.

The mapping project for the main coastal marine ecosystems is an elementary first step. But everything is complicated. The aerial photographs, the satellite images, and the pictures obtained by sonar are insufficient to make adequate maps; we must turn to diving and underwater photography to interpret the signals gathered from a distance. Thus, on the eve of the third millennium, although we know the hidden face of the moon in detail, the mapping of marine vegetation along our coasts is not nearly finished. Consequently, observations of the impact of pollution remain very localized.

In France, a system of regular monitoring of the main Mediterranean ecosystem (the *Posidonia* meadows) has nevertheless been initiated.[23] Precise mapping of underwater beds of these flowering plants is conducted every two years on the same thirty sites. The stability, retreat, or advance of *Posidonia* is a good indicator of the general evolution of the coastal environment. But the forces of change in these meadows remain largely hypothetical.

For the North Sea, researchers have established an original method, the "AMOEBA approach," designed to evaluate ecosystem health.[24] Sixty animal and plant species were carefully selected to represent the ensemble of marine habitats of the Dutch coast. The ecologists then estimated the ideal number of each species in an ecosystem in equilibrium by observing undamaged habitats or analyzing data gathered before these habitats were subjected to human activities. They thus established the characteristics of an ideal, or healthy, ecosystem that can be compared to the current state. For seals or the prairies of the marine plant *Zostera*, the registered decrease is significant, indicative of a very disturbed ecosystem. In exactly the opposite fashion, the habitat change

favors the proliferation of algae that can capitalize on the nutrients in
the fertilizer and phosphates of agricultural origin. Based on this knowledge, the authorities attempt to intervene in matters of coastal development, predation, or pollution, in the hope of re-equilibrating the populations of one or several species. The price to pay for rehabilitating a
habitat or reestablishing the population of a target species can be exorbitant. In the North Sea, the costs to assure the return of seal populations
to the Frisian Islands will be very high.

The consequences of human activity on the diversity of life depend
on the potential for reversal and the relative scale of the damage (per-
centage of an ecosystem affected in a region, or degree of destruction of
the population of a species). The degree of reversibility can be defined
as the length of time the impact persists after the damaging agent is
halted. The perturbation of the habitat can be interrupted suddenly; the
disturbance may have been a temporary mishap (grounding of an oil
tanker, toxic substance spill), or there can be a voluntary reduction (industrial or urban pollutants). If the spilled chemical products can accumulate in sediments and then reenter circulation, the interaction of di-
lution and chemical recombination can lower the toxicity level from
year to year.

The potential for reversibility depends on the speed at which the
altered populations and ecosystems naturally recover. Great differences
exist among the many animal and plant species. These biological characteristics (development, growth, reproduction, dispersal, numerical
size) determine the greater or lesser speed of recolonization of a site that
has again become suitable. Some species return in months, others not
for centuries.

For each assault on marine life, it is appropriate to draw up a time
scale for reversibility after the voluntary or involuntary cessation of pollution. This step resembles the calculation of half-lives of radioactive
substances. Thus we can distinguish among impacts.

Short-term impacts are quickly effaced after the causes cease to operate. In this category are chronic diffuse bacterial pollution and acci-

dental spills of biodegradable hydrocarbons near sandy coasts. Similarly, stocks of molluscs, crustaceans, and fishes ravaged by overfishing can recover rapidly after the complete cessation of harvest.

Intermediate-term impacts are primarily generated by non-biodegradable chemical pollutants. Contrary to widespread opinion, chemical pollutants rarely attain levels sufficient to cause the extinction of species or ecosystems. This type of change often remains localized around the point of introduction. In these situations, high concentrations of pollutants can, obviously, disrupt the populations of certain species. The most susceptible species are predators that accumulate and concentrate pollutants. But when the pollution ceases, populations and ecosystems often recover in an intermediate time span.

Long-term impacts especially affect ecosystems or species that inherently recover slowly. It requires several millennia for *Posidonia* meadows to retake one hundred meters of territory at their lower depth limit. However, their destruction is brought about by causes that at first seem minor, for example, a prolonged increase in water turbidity (produced by spillage of mud, urban water laden with suspended matter, etc.) or the repeated activities of fishing gear dragged along the bottom (dredges or trawls). The siltation and accumulation of organic matter or nutrients (nitrates, phosphates) on the sea floor also lead to long-term impacts on coastal habitats.

There are also irreversible impacts, those that correspond to the definitive destruction of a species or an ecosystem. Known cases of human-induced extinction of marine species are still rare. In the Mediterranean, only two causes are known to have led to the extinction of species or of habitats or at least to have threatened them with extinction. The first is excessive depredation. The Mediterranean seal has been nearly eliminated by fishermen because this marine, fish-eating mammal is considered a competitor. The second cause is coastal habitat conversion. Construction achieved at the expense of the sea constitutes an irreversible impact. Coastal ecosystems in effect form narrow strips, parallel to the coast, that are veritable oases of life. The attached marine vegeta-

tion, which is associated with an exceedingly diverse fauna, can live only
in this zone with adequate light penetration. Beyond this coastal zone is
a virtual desert: a muddy slope without light, absent of plant life. The
construction of a harbor or a parking lot on reclaimed coastal territory
buries these oases. The damage is irreversible: it is the amputation of a
critical space where life is concentrated.

What about the damage that *Caulerpa taxifolia* has inflicted on
marine biodiversity? The invasive and dominating character of the alga
allows us to speak of biological pollution. This is a poorly known concept, the effects of which are underestimated because it appears natural.
The observed impacts on coastal species and their habitats are much
more severe than those attributable to more traditional forms of pollution that are so routinely lamented at the beginning of every summer.
Above all, this biological pollution has become uncontrollable, and, contrary to other forms of pollution, the unpredictable endpoint of the process is dictated by autonomous living organisms.

Let us compare the impact of the *Caulerpa taxifolia* invasion to
a well-known impact on the marine environment, pollution from oil
tanker wrecks. Contrary to widespread belief, the harmful effects of
hydrocarbons spilled into the sea are always limited in time and space.
The injurious effects, though spectacular, disappear in several years. In
most cases, the affected species, which live at the interface of water and
air, can recolonize rapidly. In April 1991, the oil tanker *Haven* ran
aground off Genoa, and part of its cargo spilled. The catastrophe out-
raged Mediterranean nature lovers. People spoke of the imminent death
of the Mediterranean. At the same time, 150 kilometers west of Genoa,
Caulerpa taxifolia occupied an area of thirty hectares. Three years later,
outside the immediate site of the grounding, there was no trace of hydrocarbons on the shores of the Côte d'Azur and the Italian Riviera. By
contrast, the alga had invaded 1,500 hectares between the surface and
thirty-five meters deep.

Biological pollution propagates itself from year to year, while hydrocarbon pollution is gradually reduced. Imagine for a moment that

256   oil slicks "multiplied" from place to place and that there was no way to foresee the end of the process. Or let us use another image. In this case—a sudden spill of hydrocarbons into the sea—the damage inflicted on marine ecosystems can be compared to an ugly, brutal facial skin infection: it is visible, ugly, and distressing. But the cure will be rapid, and the skin will soon recover its previous appearance without a trace of the infection and without scars. By contrast, the invasion of a habitat by an introduced alga can be analogized to a more insidious pathology that remains unnoticed and seems harmless because it is "natural." Hidden in the body, it regularly propagates itself and runs rampant. The prognosis is unfavorable.

The three questions that we asked by 1990 with respect to this biological pollutant, *Caulerpa taxifolia*, remain unanswered today. Is the alga going to spread inexorably along all the coasts of the Mediterranean and other seas? Is it going to change food chains to the point that species we eat are affected? What will be the scope of the disturbance of coastal marine ecosystems?

We should fear the worst. We base our pessimistic predictions on current knowledge of the phenomenon, on the biology of *Caulerpa* and of Mediterranean ecosystems. But, in the name of scientific veracity, it is appropriate to admit the existence of doubt and controversy. The use of the conditional tense is always necessary, as is the use of qualifying terms for threat and risk.

Can we hope for favorable changes in the behavior of this alga? For some reason difficult to imagine today, might it disappear and leave in its wake a negligible impact? Should these hopes, if they have any foundation at all, lead us to be passive? Some people figure that we can accommodate to a lower biodiversity, while others emphasize the imminence of a global climate change that, in any event, will change all the natural equilibria that we know. Should we admit that this fate is inevitable? Time will tell who is right. For my part, I prefer to apply the principles of responsibility and precaution in the face of such a threat.

Caulerpa taxifolia is just one dramatic example of an accelerating phenomenon, the homogenization of the biosphere by species introduced
to every continent and island. The consequences of these invasions are
found in all habitats. Inadvertently or deliberately, humans have always
carried species from one region to another and, ultimately, between con-
tinents. The development of rapid means of transportation has greatly
increased the frequency of such introductions.

Most introduced species disappear rapidly from their new homes
because either the physical environment or the biotic one (predators,
competitors, parasites, pathogens) does not allow them to persist.
Among those that manage to establish populations, some do so only because humans continue to aid them (as, for example, most horticultural
varieties); these remain localized. Those that invade different habitats
and grow without human assistance are called "invasive" or "natu-
ralized."

Many of the first introduced species were agricultural plants and
animals. The majority of the most heavily consumed species in both
Europe and North America are not native to the regions where they
are now heavily cultivated. Other early introduced species were those
preadapted to hitchhike with human travel and commerce. For example, most of the first insects introduced to North America were European beetles that lived in the soil used as ballast in sailing vessels travel-
ing to North America to return full of raw materials from the New
World.[25]

The first alien species recognized as pests were mostly those that
ravaged agriculture, especially insects and plant pathogens. A legendary
early example is the devastating invasion of European grape vines by a
tiny insect.[26] In 1860, new winemaking methods had recently allowed
wines to be stored for long periods, a development that brought about
an enormous increase in land devoted to viticulture and in research on
cultivated varieties of grape vines. An agricultural engineer from the

258   Bordeaux region imported vines from the United States; the roots contained a small sap-sucking insect, the phylloxera. The American vines tolerated this insect well, but the European ones did not. In less than twenty years, the phylloxera spread from region to region destroying the majority of European vineyards, sowing ruin in the agricultural community and forcing many to emigrate to North Africa. More than a million hectares of vineyards were destroyed. More than 5,000 patents, the majority far-fetched, were granted for methods to battle the invader. The French Academy of Sciences, then presided over by Louis Pasteur, was engaged to solve the problem. Finally, Louis Planchon, professor of botany at the University of Montpellier, recognized that it was the phylloxera that was causing the vines to die and found a solution—importation of American rootstock, onto which the French vines could be grafted. American rootstock was massively imported between 1880 and 1900. Another invader was thus accidentally introduced, a mildew. This fungus was a new calamity for vineyards that had barely been reestablished. To battle this invasion, wine growers had to spread a copper solution at particular times of the year. Today, a century later, these two invaders are well under control; all the rootstocks come from American vines and Bordeaux mixture (a copper solution) is regularly applied to the vines. Consumers of wine do not realize that every time they enjoy a glass of burgundy, it is thanks to the continuing control of these invasive pests, phylloxera and mildew.

Hundreds of other imported insects and fungi continue to threaten our main crops, necessitating a major, endless battle. About a third of all crop production is estimated to be lost to introduced pests nowadays.[27] A large part of modern intensive agriculture is dedicated to the control of invaders—use of resistant varieties, use of fungicides, herbicides, and insecticides, development of biological control, and most recently the creation of transgenic plants resistant to certain pests.

Many introduced species have invaded natural habitats to the detriment of one or more native species. Aside from economic consequences of varying degree (including loss of recreation and tourism), biodiversity is threatened. A hierarchy of impacts on biodiversity can be erected.

In each hierarchical level, the gravity of the case depends on the vigor of the invader, its dominance, its rate of spread, and its persistence.

At the first level, the introduced species maintains itself in a limited range of habitats without spreading and without upsetting the equilibrium of the ecosystem. The species thus occupies an "empty" ecological niche. This situation allows two interpretations. First, one can see this introduction as an alteration of the ecosystem by an alien element that at least modifies the species composition, even if it appears innocuous otherwise. Second, one can, by contrast, see this introduction as beneficial because it has enhanced local biodiversity.

At the second level, the introduced species spreads to the detriment of one or a few native species. It thus threatens native biodiversity. Many examples are known. In Europe, the red-eared slider *(Trachemys scripta elegans)*, a temperate North American turtle of which very young individuals were imported for private aquaria, was released in nature, where it matures to up to 25 centimeters and becomes voracious. Oddly, the French call this the "Florida turtle," even though it was first introduced to Florida around 1958.[28] In Europe, it is gradually eliminating the native freshwater turtle *Emys orbicularis*. This interspecific competition is the sole impact of the Florida turtle on biodiversity. The eastern North American gray squirrel *(Sciurus carolinensis)* was introduced to Great Britain beginning in 1876. It has spread widely and outcompeted the native red squirrel *(Sciurus vulgaris)*, particularly in deciduous woodlands and manmade habitats.[29] Populations of the native species have continued to decline, but this is the only major impact of the gray squirrel documented in Great Britain. The fiercely predaceous brown tree snake *(Boiga irregularis)*, a native of eastern Indonesia, the Solomon Islands, and northern Australia, arrived in Guam in military traffic around 1950. It gradually spread from the port until it was common throughout the island by 1982. Nine or ten of the twelve native forest birds have already been driven to extinction by the snake, and the remaining two species are rare.[30] Impacts on other species are undocumented.

At the third level, the introduced species becomes dominant and alters or upsets the entire ecosystem. Again there are numerous examples. I will cite two of the best known.

A comb jellyfish (ctenophore) with the strange name *Mnemiopsis leidyi* looks like a small, translucent medusa. It has caused one of the most dramatic and damaging invasions of the last quarter of this century.[31] It is native in the estuaries along the western Atlantic coast from the northern United States to the Valdés peninsula in Argentina. It was almost certainly introduced by a ship that loaded *Mnemiopsis*-laden ballast water in the western Atlantic and then emptied its tanks in the Black Sea, where the comb jellyfish was first detected in 1982. The ctenophore was first misidentified, and not until 1989 was it recognized as a species of *Mnemiopsis* and thus an invader. It usually has moderate population densities in the western Atlantic, but the populations exploded in the Black Sea and the adjacent Azov Sea and Sea of Marmara. This example demonstrates the enormous impact that a small, apparently innocuous species can have in this habitat. This species has invaded the entire Black Sea, a practically closed sea that communicates with the Sea of Marmara and thus the Mediterranean through the Turkish Straits of Bosporus. The Black Sea has two unusual features. On the one hand, it is naturally sterile at great depths; there is no oxygen between 200 meters and the deepest regions, which surpass 2,000 meters. On the other hand, it is highly polluted, as it receives the great rivers of eastern Europe and Russia, which drain the polluted effluent of many giant factories and large cities with inadequate sewage treatment. Thus, the quantities of nutrients, insecticides, fungicides, herbicides, heavy metals, organic compounds, hydrocarbon derivatives, and radioactive waste found on the edges of the Black Sea near the estuaries of the great rivers are all worthy of mention in the *Guinness Book of Records*. Despite this unenviable situation, which would not seem conducive to life, the catch of pelagic fishes (primarily anchovy, sprat, and horse mackerel) had always been good. But when *Mnemiopsis* exploded in 1988 (more than 1 kilogram/square meter, up to 500 individuals per cubic meter) and devoured all the zooplankton including fish larvae, the entire pelagic ecosystem was profoundly modified, and the catch plummeted. The anchovy catch fell from 204,000 tons in 1984 to 200 tons in 1993; sprat from 24,600 tons in 1984 to 12,000 tons in 1993; horse mackerel from 4,000 tons in 1984 to zero in 1993. A simple little comb jellyfish caused more damage to the fishery than the various pollutants so often decried! As its food base declined, the *Mnemiopsis* population began to collapse in 1991, but the ctenophore is still present, with drastic annual population fluctuations. Though we can reasonably hope for a reduction in pollution from the Danube, Dnieper, Don, and Dniester Rivers, what can we hope to do against *Mnemiopsis*, which has overthrown the entire pelagic ecosystem of the Black Sea?

The prickly pear cactus *(Opuntia stricta)* probably holds the record for area infested by a terrestrial invader. A single *Opuntia stricta* plant imported from South America to Australia in 1839 initiated an invasion that reached more than 24 million hectares of arid scrubland and rangeland. The invasion was largely contained and vast areas cleared of the cactus in a triumph of biological control. Caterpillars of a small South American moth, *Cactoblastis cactorum,* feed exclusively on the cactus in its native range. Imported beginning in 1925, they reduced prickly pear to a small, ephemeral problem. Both cactus and moth remain in low densities in Australia.[32]

At the fourth level, the introduced species affects several ecosystems, thus threatening an even larger swath of biodiversity. The number of invaders of this sort is growing. They are, for the most part, species able to tolerate a wide variety of habitats, or those in such great densities that they disturb all the ecosystems surrounding the one they inhabit.

Water hyacinth *(Eichhornia crassipes)* is one of the most widespread invaders worldwide. A century after its first introduction outside its native range, the Amazon basin, it infests numerous tropical lakes, estuaries, streams, and rivers. A beautiful plant that attracted botanists seeking ornamentals for botanical gardens, it was imported to a horticultural exposition in New Orleans in 1884. Visitors were impressed by its beauty and planted it in several water bodies. The aquatic ecosystems

262   of the southeastern United States were then progressively colonized by
vast, floating, dense carpets of water hyacinth. The economic repercussions (especially interference with navigation) first drew attention, but
the presence of an opaque covering of plants on the water surface and
the eventual decomposition of dying plants devastated numerous
aquatic ecosystems, both planktonic and on the bottom.[33] At one time
50,000 hectares of Florida waters were dominated by water hyacinth.
There it has been reduced to a minor problem, primarily by the use of
chemicals and large floating mechanical reapers, but the plant remains
a pest in many states, particularly Louisiana. Water hyacinth reached
Africa in 1892, then Asia in 1894 (after being brought to a botanical
garden in Indonesia). Today water hyacinth is present around the globe
on thousands of kilometers of streams and rivers. It first appeared in
great quantity in Lake Victoria in 1989. Today it covers over 5,000 hectares and is spreading. It wreaks havoc with the commercial fishery, fouls
boat engines and propellers, obstructs landing sites, and clogs cooling
pipes for power plants, leading to massive blackouts.[34] The impact on
native species must be enormous but is largely unstudied. This insufficient scientific documentation of ecological impact is lamentably common for most ecosystems invaded by this plant.

Other floating aquatic plants introduced in various regions have
caused similar damage to freshwater ecosystems but are not as widespread geographically. The arrival in African and northern Australian
estuaries, plus some Asian areas, of the South American water fern *Sal-
vinia molesta* is one of the worst aquatic weed invasions.[35] The first escape of the fern was from a botanical garden in Sri Lanka. It is not
known to reproduce sexually, but its vegetative reproduction is prodi-
gious. Other examples of invasive floating plants are more regional—
Eurasian and South American water milfoil (*Myriophyllum spicatum*
and *Myriophyllum aquaticum*, respectively) in lakes and streams of the
United States, Canada, Australia, New Zealand, and South Africa; *Elo-
dea canadensis* (Canadian pondweed) in northern Europe and Australia;
tropical water lettuce *(Pistia stratiotes)* and Asian hydrilla *(Hydrilla
verticillata)* in the United States.[36]

The invasion of North American lakes and rivers by the zebra mussel *(Dreissena polymorpha)* is another level four case.[37] These small
mussels (less than five centimeters) from central Europe (the shores of
the Caspian Sea) were introduced to the canals and rivers of Great Britain as early as 1824. The first individuals observed in the Great Lakes
of North America were discovered in 1988 in Lake Saint Clair, but they
had probably arrived two to three years earlier. The zebra mussel has
spread dramatically and continues to do so. It first infested the Great
Lakes, then the waters communicating with this system, reaching the
Mississippi River and as far south as Louisiana and Tennessee. It is now
in at least nineteen American states and two Canadian provinces. The
mussel tends to cover solid substrates, but it has also settled on soft bottoms. Concentrations are often staggering—up to 900,000 individuals
per square meter and layers of shells up to fifteen centimeters deep!
This colonization has drastically modified benthic ecosystems; the dominance and extreme density of mussels on rocks eliminates many sessile
native species and seals off many shelters. The billions of mussels filter
enormous quantities of water, to the extent that the water becomes far
more transparent. This change favors the colonization of the bottom by
rooted plants. The phytoplankton that the zebra mussels eat decline in
density; the zooplankton that eat the phytoplankton, and the trophic
web associated with the zooplankton (including fishes)—the invasion
affects all of them. In addition to the ecological impact, the economic
impact is staggering. The zebra mussel clogs water pipes of power plants,
water treatment facilities, boat motors, factories, and other structures.
This impact has been evaluated at $3.1 billion over ten years for the
electric power industry alone![38] There is currently no known impedi-
ment to the further spread of the zebra mussel.

Caulerpa taxifolia, a dominant, persistent, rapidly spreading introduced species, is ubiquitous—it colonizes diverse habitats. A variety of
habitats are thus affected by this invasion; it falls squarely in level four,
the highest in the degree of threat to biodiversity. The fact that it appears to be a single individual, a clone, of a genotype unknown in nature
makes it an exceptional case.

Although introduced species are found everywhere, some regions are more heavily invaded than others. Australia, Florida, and the Hawaiian islands are among the regions suffering the most upheaval from invasive introduced species (in Florida, twenty-five percent of the species found in natural habitats are introduced).[39] They are also characterized by many endemic species, so the impact of the invaders on biodiversity is exacerbated.

In the United States, more than 7,000 introduced species (not counting microorganisms) are established in nature, of which perhaps fifteen percent cause ecological or economic damage.[40] Some recent cases are rapidly evolving. The cordgrass *Spartina alterniflora* of the Atlantic coast of the United States has massively invaded soft-bottom coasts of California and Washington, completely transforming intertidal ecosystems. Kudzu *(Pueraria montana)*, a Chinese vine, has spread through the forests of the Southeast and Hawaii, covering more than 1.6 million hectares with a green curtain. The European green crab *(Carcinus maenas)* is invading the Pacific coast (and also Tasmania) in enormous numbers, with major impacts on coastal benthic food webs.

Each invading species is a unique case, with characteristic impacts, degrees of dominance, and features of dispersal. Thus each invasion has been treated differently. But the succession of invasions, each dramatic in its own way, that spreads rabbits, rats, camels, horses, deer, birds, frogs, toads, snakes, fishes, insects, jellyfish, crustaceans, molluscs, starfish, sea urchins, dinoflagellates, macroalgae, ferns, and higher plants is dizzying. The atlas of plant and animal pests expands continuously. Legislation to slow this flow is drastic in a few nations, but rare or nonexistent in the majority. The scientific illiteracy with respect to the global threat posed by invasive introduced species means that other ecological horrors are much more in the news. Insidious (because it seems natural), progressive, underestimated—this is the blow struck against biodiversity by the flood of human-introduced species. Has it not already surpassed that caused by the sum of all chemical pollution?

The Decline of Ecology

The decline of ecology is reflected by lack of respect for professionals in the study of life in its natural habitats, ecologists. The various episodes that have marked the story of *Caulerpa taxifolia* well illustrate the difficulties encountered by ecologists, specialists in this domain, in trying to make their opinions heard by society or at least by its official representatives.

In this affair, the scientific abilities required to justify the alert were many. It was necessary first to verify that the alga was indeed a newcomer to the Mediterranean, and for that one needed a thorough knowledge of the Mediterranean marine flora. To identify the species, knowledge of algae in the same group, censused in all the tropical seas, had to be mastered. The recognition that it tended to colonize Mediterranean bottom patches had to be founded on the observation of the alga in the field (in this case, underwater). To be able to predict and explain its spread, solid knowledge of its biological characteristics, its reproduction, its growth, and its development was indispensable. Finally, to estimate the risks associated with its proliferation required good knowledge of Mediterranean ecosystems.

In one sense, these diverse facets of knowledge closely match the profile of a professional naturalist, a specialist in a very narrow field. However, good knowledge of Mediterranean ecosystems requires broad knowledge of many specific disciplines. The observation of an invasion by an introduced alga and the definition of the associated risks constituted a textbook case for a marine ecologist. Possessing this knowledge, such a person should have been able to occupy the role of the sentinel who gives the alert after identifying a danger. Once the alert is given, the relevant authorities should have trusted this person. But they did not believe in the existence of a major environmental threat. They should have been sensitive to the arguments that were advanced, for they were very simple, and all the while they could have remained critical. But they did not even bother to attempt to verify naturalists' state-

266 ments. Because the problem persisted, they could have reversed their position or launched an assessment. But they preferred to maintain their stance, choosing the strategy of "wait and see."

Can we say that the extraordinary complexity of life justifies the ignorance, indeed the contempt, of the authorities regarding this affair? Can uncertainty about how a biological phenomenon will unfold justify the passivity of competent authorities? Do either the memory of other cases of fleeting algal invasions, with minor impacts, or the fact that many other marine organisms have been transported by humans permit one to issue reassuring statements? I remain certain that this phenomenon presented all the traits of a classical catastrophic invasion, similar to those described in most ecology textbooks.

Let us recall the main facts known to us by 1990. Never observed before 1984 in the Mediterranean, the alga definitely had to be considered an *introduced* species; proliferating under the Museum of Monaco and then in France, it was certainly *invasive;* knowledge of its biology surely allowed the prediction that it would *persist;* in competition with other algae, it proved to be a *scourge* capable of disturbing ecosystems, biodiversity, and eventually our marine natural resources. We should add to this gloomy picture an aggravating characteristic, that of the suspected presence, later confirmed, of toxins. In short, it was legitimate, urgent, and of the highest priority to concern ourselves with this alga, a threat to the Mediterranean environment.

It is too easy to invoke the problems of personal rivalries, of competition, of scientific power, or of internecine laboratory strife to explain the hesitation in confronting this ecological threat. Cumulative effects of negligence in the political evaluation of the problem distorted the scientific assessment and favored inaction. The best government experts in the nation were unable to identify and manage a serious attack on biodiversity. Beyond the lack of my credibility, it was in fact the entire discipline that was not recognized as trustworthy. This affair demonstrates how the knowledge of life and its habitats is the victim of lack of understanding, clumsiness, and incompetence, not to speak of lack of

interest and even disdain. This attitude is translated into the overall decline in instruction and research in the sciences dealing with nature.

REDUCTIONIST BIOLOGY VERSUS THE SCIENCES DEALING WITH NATURE

The sciences of nature cover a wide range of interdependent themes, including ecology, the biology of individuals and populations, ecosystem dynamics, and taxonomy or systematics (the science associated with identifying species).

In France, the situation is grotesque. Among the elites who direct institutes in charge of the marine environment, very few have pursued suitable university studies. Here one finds a geographer with a literary bent, there an otorhinolaryngolist, elsewhere a highly skilled trade union economist. The highest bureaucrats responsible for dealing with our environment are products of the "grandes écoles,"[41] in which instruction in marine ecology does not weigh heavily in the courses of study. Whatever other merits they may have, these persons in charge of protecting the sea have never had scientific careers. Neither would Commander Cousteau and Dr. Bombard,[42] symbols of marine conservation, claim to have been career scientists. It would be difficult to imagine a comparable situation in medicine—a chiropractor, a geologist, or a poet appointed by the authorities to head the great medical institutes, becoming spokespersons for the health professionals.

This situation reflects the weakness of disciplines dealing with the study of nature in the field. Productive organizers of research teams are rare and increasingly marginalized by their colleagues in biology. In effect, the root cause of the decline of ecology has its basis in the devaluing of the sciences of nature within the overall discipline in which it is embedded, biology. In one generation, biology has completely changed. It is now almost entirely devoted to the study of the microscopic: cells, their functions, molecules, and most recently genes. From physiology, we have progressively moved towards molecular biology. The current generation of biological researchers is composed essentially of labora-

268 tory scientists. Their field of expertise has been reduced to a single level, the molecular. This approach to the study of life is termed "reductionist."

The reductionist biologists, modern, progressive, transformed into chemists—chemists of organic molecules—reign over biology. I do not in the least underestimate the importance of the discoveries in this domain and the unexpected applications they bring to medicine and agriculture. But I regret the imperialism of molecular biologists, which has had the effect of reducing nearly to silence the practitioners of taxonomy and the other sciences of nature perceived nowadays as old fashioned. Because there is not room or funding for everybody, this overwhelming dominance of molecular biology has led to a reduction of positions and grants for the other disciplines.

I am neither the only nor the first to point to the rout of ecology. The greatest and most highly respected ecologists reiterate it endlessly: the status of ecology in France is a disaster, and a dramatic one.[43] These clear-sighted naturalists accuse the scientific decision-makers and the reductionists of ostracizing them, of sectarianism, of dogmatism, of irrationality, and even of being illiterate with respect to nature. They denounce the aberrations and errors published by the reductionists in power, removed for too long from their base—life in the field, the behavior of species, and the classification of organisms.

THE PROCESS OF MARGINALIZING ECOLOGY

"For every kind of dominant ideology, two critical analyses can be opposed. One stigmatizing the more or less pernicious nature of the 'vision of the world' that forms this ideology; the other, the very exercise and abusive methods of its domination. It is on these two levels that one should denounce the violence of the system," wrote François Brune.[44] Among the causes of the collapse of ecology, we must first invoke the immense gap between the amount of research funding devoted to reductionist biology and the amount granted to the sciences of nature. Certainly, naturalists can make their laboratories function with a smaller overall investment. Because the management of the environment is the responsibility of the public sector, naturalists are supported only by the national government or community groups. In France, the recent economic crisis has led to a noteworthy reduction in institutional resources, even when they have not dried up entirely.

Moreover, here and there, ecologists undertake studies concerning habitat conversion or pollution. Their professional opinions are not always well received. It is well known that certain teams have been accused of thwarting some huge project or branch of industry. Blacklisted by the promoters and their political allies, they have seen their funding decline drastically. Ecologists experience a sense of urgency to define an ethical framework in which their responsibilities will not be questioned. More than other scientists, they are confronted directly with corruption when they are solicited to defend questionable projects. Passivity is the most widespread response: don't say anything, don't criticize an attack on the environment in one's area of competence, in order to preserve the means to keep functioning.

The situation for reductionist biologists is an entirely different matter. Their disciplines are open to the private sector, that of the pharmaceutical and biotechnology industries. Grants and encouragement from the national government have fostered enormous investments and have allowed many laboratories to take off. Patents and licenses for development assure dedicated funds for certain laboratories. Finally, large philanthropic organizations gather enormous sums to offset the possible funding deficiencies in the system or to enhance the means of the wealthiest. Of course, abundant, regular funding does not suffice to ensure supremacy; one must also garner positions, jobs. It is on these grounds that the reductionist biologists have acquired and enhanced their preeminence. They have succeeded in imposing their rules of the game and their conception of "research biologist" in all recruitment proceedings.

It is certainly at this level that the conflict between biologists is most pernicious. In imposing their own rules, the reductionists have devalued the intellectual accomplishments of naturalists and defined to

270   their own benefit the rules for evaluating the productivity of a research biologist. How did they accomplish this? An enzyme, a molecule of a designated species, possesses a universal formula and invariant modes of action. Whether they are in Canada or Australia, researchers are confronted with the same logic: be the first to describe its function. A sole means of communication is legitimate: publication at an international level in an English-language scientific journal. The rule is simple and precise: the score given to a researcher, laboratory, or institution depends on the number of publications. To decide between competitors, there exists a more elaborate classification that gives bonus points for publications in certain journals, those that are cited the most and therefore are the most valued. Because they are by far the most numerous researchers in biology, the reductionists have at their disposal twenty international journals dealing with all areas of molecular biology. The small number of journals insures that each is cited very often, and therefore is highly valued.

By contrast, the naturalists are dispersed in numerous, independent micro-specialties. For example, for vertebrate ecology or biology, there are specialists in fishes (ichthyologists), in birds (ornithologists), in snakes, frogs, and lizards (herpetologists), and in mammals (mammalogists). Each micro-specialty is covered by several international journals, so the circulation of each journal is low, and its citation rate is minimal outside the small circle of specialists. Thus the hit parade of biological journals is occupied by molecular biology journals.

This tendency not only devalues the production of knowledge by naturalists, but it is accompanied by a retarded development of their research and their careers. For several decades the reductionist disciplines have monopolized not only a good part of the research funding but also many financial rewards from industry. The most prestigious among these rewards is the Nobel Prize in medicine and physiology, which, as its title indicates, does not deal with ecology. The award of the Nobel Prize to André Lwoff, François Jacob, and Jacques Monod in 1965, all three of whom were molecular biologists, greatly favored the development of this discipline in France.

Among the causes of the decline of ecology, we should finally emphasize the great disparity in constraints on the organization of research between the two types of biology. Like computer science, the study of the molecules of life is a new science; the bibliography is limited in time, thus easily available thanks to the computerization of all the data. Moreover, molecular biology is such a fast-moving field that after ten years the majority of writings become obsolete, if not downright wrong. The investment required for archiving the data and bibliographic research is thus minimal. Another feature of this scientific domain is the necessary modernity of the laboratory, which assumes there will be many users to defray or justify the investment. The work is collective, it allows regular publication and registration of patents, two things that lead to honors and that, as a further indirect result, facilitate renewed funding. The only certifiably high-quality (in more hypocritical terms, one would say "recognized") teams are those comprising at least ten scientists conducting research on a (unique) unifying theme. Thus in a good laboratory, a beginner surrounded by the right people can, after six months of initiation, co-author his or her first international paper.

By contrast, naturalists have to master a heritage of several centuries of knowledge. Using old writings and managing collections of specimens or herbaria confers on them the role of archivist or museum researcher, activities with pejorative connotations for other biologists. To flourish, one must specialize. Each person's research is thus oriented towards a range of knowledge limited to several ecosystems or a group of species. A naturalist who is productive will be able to become one of the ten internationally recognized authorities in a field. Everything that has been published in the area in the last two centuries ought to be, if not known, at least accessible. To understand the functioning of an ecosystem of a region and to predict its evolution, one must begin by compiling and assimilating all available data. One must also call on other disciplines, like molecular biology or genetics, but also geology, meteorology, satellite imagery, or even sociology. Ecology deals with very broad concepts; it is said to be "holistic."[45] Because a single researcher cannot simultaneously excel in systematics, in knowledge of an environment, in

272   the techniques of biochemistry or satellite imagery, the strength of a laboratory of naturalists resides in the collaboration of complementary specialists.

Consider the example of a specialist in plant ecology.[46] First of all, he or she will not be able to aspire to be qualified until more than five years of introduction to botany, by studying old books and herbarium specimens. He or she will then be able to initiate a study, for example the analysis of the evolution of the flora of small valleys subjected to human change. This will require two years trudging around in the field to gain the detailed knowledge of the flora that flows only with the passage of time. He or she will call on other specialists (chemists, geologists, geographers) for better knowledge of the causes of environmental changes or better evaluation of the areas occupied by the delineated plant communities. He or she will record in detail what has been observed or deduced. This contribution will be appreciated by the managers of the site that was studied, who were able to provide the researcher at least with logistical support, and at best with modest financing of a laboratory. This work will have a historical value—the status of the flora at the time the data were gathered will have been captured. But the thick report, difficult to publish, will remain private. It can be condensed and published in a local or at best national journal, but that will hardly increase the author's standing in the eyes of decision-making authorities and reductionist colleagues.

The ecologist is a pinochle-player in a circle of bridge-players. However qualified he or she is, however good his or her reasoning, however exact and well presented his or her results, a publication in French in a national journal has little value—next to nothing according to the criteria for evaluating the "modern" biological researcher. If he or she ventures to publish a review in English in an international journal, the journal chosen will at best be the one most quoted in the specialty. Wasted effort—it will not rank among the elite journals, all dominated by reductionists!

The years of collecting specimens, observations, accumulating data of all sorts will, however, make the ecologist or systematist a priceless expert for those in charge of managing our natural areas. Thus the naturalist participates, as an ex-officio university expert, in scientific committees for protected areas (national parks, regional natural parks, nature reserves). His or her opinion on pressing questions about the management of the environment is solicited by the prefecture or the general council. A naturalist is available as a volunteer. But none of these professional service activities are recognized in decisions on career advancement.

Facilely accused of frittering away time in micro-specialties without direct application, of wasting effort in committee duties on environmental management, of being too independent or uncontrollable and above all unproductive with respect to top-ranked publications, the naturalist is thus depreciated by his or her supervisors. This is how the reductionists corralled the majority of positions in biology.

Reorganize! Unite around great themes! Find unifying themes requiring expensive equipment! Research administrators and assessment teams echo these principles that underlie reductionist biology and other scientific disciplines like physics and chemistry. This is the carrot used to obtain facilities, positions, grants. It is as if a group of prosperous house painters wanted to convince needy artists who have fallen from favor to paint one wall as a team, in a uniform style, with the best equipment!

But the message is even more blunt than this. The real mandate is to stop these worthless, individual flights of fancy of descriptive biology.[47] The admonitions of a well-known biologist run along the same lines: "Reductionism represents a necessary phase for every experimental science. Sooner or later, the biologist cannot escape from this rule. It is very necessary that the naturalistic, contemplative attitude of the first systematists and comparative biologists change, in step with scientific and technological progress, into a 'precise analytic mode.'"[48]

Molecules will henceforth be the sole basis for all of biology. Between the two branches of biology, the abyss of lack of understanding deepens. Faced with repugnant ostracism by some reductionists, those collectors of English-language publications, how many botanists, zoolo-

274 gists, and ecologists have chosen to abandon their research? How many great specialists, rejected by the "system" and subjected to the constraints of idleness, have been discouraged or have fallen into the depths of depression?

In the largest and most prestigious French research institute, the CNRS, some ecological colleagues worry about whether the word "ecology" has become taboo! In 1995, the department of life sciences of CNRS comprised 335 research units, of which only 49 focused on nature; among these, more than half approached the subject by molecular techniques. In the whole of France, only five to seven specialists in the study of nature are recruited every year by this organization; the last few years, their degree of collaboration with reductionists has determined who is chosen.[49] Students excited by biodiversity quickly learn that to embark on this path is unlikely to lead to a satisfying future. We must admit it: our best students are turning away.

It is not through a few isolated but highlighted programs of research on the environment that a nation can hope to maintain a discipline so necessary to the management of its environment. The development of several dynamic ecological laboratories will not hide the profound misery of dozens of laboratories of botany, zoology, and ecology that are waning to extinction in our regional universities. The repair— in emergency situations—of the great hall of evolution of the National Museum of Natural History in Paris will not cause us to forget the dilapidation of our less visible collections, like herbaria. The new "environmental generalists" produced by the ream in short instructional programs (one or two years) will not replace specialists.

Sometimes hope returns with the emergence of global environmental problems, like the greenhouse effect or the ozone hole. This hope is quickly dashed; research programs on the environment, initiated with great fanfare, are usually supervised by molecular biologists. In association with chemists or physicists (to study the atmosphere, for example), molecular biologists have made the classical reductionist research themes more "green"; one studies biodiversity or the effects of pollution at the molecular level, one models biogeochemical cycles, one classifies species by their genes. Future biologists will inherit the shell of a field (accessory sciences) and a laboratory career, which they will spend in front of computers and test tubes (basic research).

Environmental physiology, molecular ecology, genetic taxonomy, to turn to popular themes, are certainly exciting and worthy areas of research. But locked into theoretical models, the ecologist who has become a reductionist remains dependent on an analytical approach that will surely not allow a rapid assessment of environmental threats and strategies to manage them. The current trend is away from the synthetic approach necessary to decipher the many possible entrées to a complex world.[50] The intelligence born of common sense, gained from field experience and direct observation of nature, is being lost.

To complicate matters, any militant who defends a patch of land or a mistreated animal is called an ecologist or environmentalist. Environmentalism, that is, political ecology, adds to the terminological mix and exacerbates the confusion. As François Ramade has emphasized, environmentalism is a "current of thought distinct from science and, a fortiori, from research, that has provoked a very negative reaction on the part of decision-makers in general and politicians in particular."[51]

Without real institutional support, disdained by their reductionist colleagues, suspected or controlled by the national government and the regional communities, ecological researchers are consigned to oblivion. The myth of the ecological savant does not exist. In this context, we should not be surprised at the appropriation of "scientific truth" on ecological matters by powerful groups with enormous means at their disposal and masterful abilities at using the media, groups with primarily economic or political underlying objectives.

The situation in ecology is paradoxical, to say the least. While the necessity to preserve our planet and to evolve a sustainable means of development for its human population is proclaimed in all official discourse on the environment; while increasingly numerous environmental lawyers draw up legislation, issue decrees, proliferate orders of protection for species or habitats; while administrators of protected habitats are installed and granted the means to use increasingly elaborate monitoring; while the discoveries of reductionist biologists require a growing effort to assess the risks they might create (especially with liberation to the wild of genetically modified organisms), we find fewer and fewer competent ecological experts. The result is that measures of protection are increasingly empirical. Taxonomists and specialists in the study of natural equilibria are rare in parks and reserves; in scientific advisory committees for protected areas, we find more administrative representatives (enthusiastic but uninformed), reductionist biologists, and self-taught amateurs (engaged and sometimes highly competent) than true professional ecologists. Scientific personnel assigned to the study of nature, widely demobilized, are aging. Specialists on faunistic and floristic groups are not replaced after they leave or retire. They end their service before transmitting to younger colleagues their experience and their knowledge, acquired over several decades. Established schools with major specialization in ecology, formerly recognized throughout the world, have been dissolved after the departure of their most eminent leaders. It is as if clinical medical specialties like cardiology, gastroenterology, and pediatrics disappeared after the retreat of the current professors, replaced by reductionists in the same disciplines.[52]

It is not surprising, in this context, to observe that teaching biology has swung, in all programs and at all levels, towards the preponderant description of the molecules of life and their functions.[53] This evolution is not about to reverse itself. During their preparatory studies, the twenty most recently promoted teachers of "natural sciences" have received more reductionist biology than training in the science of nature.

For the first time since the rapid expansion of the exact sciences two centuries ago, a science is declining, the level of knowledge in a discipline is perceptibly dropping. Archives, documents, and collections having value as records of the historic status of nature are being destroyed. There is no certainty that there will be scientific study in this key domain in the future. It is not just extinction of several well-known species that we should regret; we should also regret the fact that we are no longer able to know the real number of these extinctions, according to a prominent American ecologist.[54] And we should add: the future historians of science will begin their descriptions of the ecological crisis of the coming decades by depicting the extinction of systematists, of naturalists, of biogeographers.

This pessimistic but realistic picture of the state of ecology explains many of the shortcomings of environmental management. It suggests that we should fear many other fiascoes in addition to the affair of the little green alga that was able to proliferate unhindered, under the very gaze of a cohort of experts whose ecological training left much to be desired. The episode of the publication of the pathetic article by the French Academy of Sciences on the so-called "metamorphosis" of *Caulerpa* is a good illustration of the illiteracy in the sciences of nature that afflicts the highest circles of biologists.[55]

Vagaries and Stakes in Scientific Communication

NO SCIENCE WITHOUT COMMUNICATION

The communication of research results is an integral part of a researcher's work. Every scientific advance, whether experimental or theoretical, is the subject of an article that is submitted to one of the many professional journals in the world. Every journal has an editorial board that calls on referees to assess the quality and originality of the article. This process is anonymous; it leads to the rejection or acceptance of the article. The scientific publication allows the researcher to present work to the scientific community, which also permits him or her to certify a date when the work was completed and to put a stamp on a discovery; it is analogous to registering a patent or a compendium in a casebook in jurisprudence.

This system is fundamental for the entire scientific establishment. It certifies the value of the science and authorizes a widescale diffusion of the knowledge gained. The reputation of a scientist, laboratory, or university, the trajectory of careers, research trends, and financing—

278 all are based on the number and quality of the publications. Quality is measured by the circulation of the journal where the research is published. "Publish or perish" is the law of survival for every scientist.

The communication of scientific results arose with modern science. It has become increasingly organized. The letters used by seventeenth-century savants to exchange information were succeeded by national specialized journals, then international journals. The majority of them are controlled by scientific societies financed by the researchers themselves and sometimes supported by government subsidies. The circulation of information is free in principle, knowledge is universal in principle, and its control appears to be democratic.

Today, the scientific literature is superabundant.[56] Whatever the nationality of the scientists, the research is most often written up in English and published by English-language journals. The journals are increasingly specialized, reflecting the atomization of knowledge into ever more specialized micro-disciplines. This ultraspecialization means that a researcher no longer has the capacity to read articles in disciplines other than his or her own and also has increasing difficulty staying abreast of progress in his or her own discipline. This tendency largely explains the growing infatuation with international multidisciplinary journals like *Lancet* in medicine or *Nature* and *Science* in the scientific domain. In these prestigious periodicals, which make a real editorial effort and reject more than 90 percent of the articles submitted because there is such demand for the limited space, short, synthetic notes are published, chosen primarily for their scientific—and media—impact. Other journals serve more to detail the experimental methods and results; they provide information more useful to specialists.

For the past several years, scientific journals have not escaped the rules of international competition that henceforth will govern the world. The scientific publishing market amounts to millions of dollars. Journals must attract the greatest number of readers and make their production profitable by any available means. This is why the large international journals have specialized in producing scientific scoops. The published research, even if it does not always constitute a remarkable advance, should be surprising. With this sensationalist rule governing their choices, editors have succeeded in attracting authors of the discoveries judged the most interesting. Their journals have imposed their values on the practice of science; they are the most widely circulated, the most widely read, and the most frequently cited worldwide. The circulation of each weekly issue of *Nature, Science,* and *Lancet* greatly surpasses 50,000 copies.

To enlarge their readership, the journals do not hesitate to turn to methods that seem more adapted to the sale of vacuum cleaners than to that of the scientific literature. *Science* and the *Annals of the New York Academy of Sciences,* for example, accompany a subscription form with a personalized pseudo-diploma. I have thus received on several occasions a diploma bearing my name, with an ornate border and authenticated by an "official" seal. This certificate is designed to flatter the recipient, who has been recognized as a researcher of high level and "elected" to membership in the honorable American Association for the Advancement of Science (AAAS, which publishes *Science*) or of the prestigious New York Academy of Sciences, on the condition, of course, of subscribing to the journals of the societies that publish them. If the publishers of English-language journals color the scientific environment in this way, it is because it enhances the income from the sales of the journals. There is something to be worried about with respect to the evolution of scientific values.

Conflicts concerning priority in publishing information have begun to appear. These last few years, researchers have observed a strengthening of publishers' rights to profit from publications, a "right" they have enforced by making authors sign a waiver of all author's rights. If a scientist wishes to republish these unremunerated writings in another journal, he or she must seek the authorization of the first publisher, who holds the exclusive rights and will not hesitate to profit from them.

These journals, termed by some "major," by others "primary," also play the role of international agencies in the matter of scientific information. They have in effect become unavoidable for scientific reporters, whether they work for a lay journal, the general press, or a radio

280   or television network. As these people have difficulty analyzing the
weighty intellectual productions of the different disciplines, they find
in *Nature, Science,* and other, similar journals excellent condensed ver-
sions of sensational news in all areas.[57] Whatever its real scientific import, a piece of news published by one of these journals acquires legitimacy in the media. This is why the majority of subjects treated by the
major newspapers, which in turn reach the general public, derive directly from the choices made by a handful of English-language scientific journals.[58]

To strengthen this tendency, the publishers of these journals have
developed an aggressive policy of self-promotion. They make available
to the media the abstracts, even certain entire articles before they appear, imposing a gag order until the date of publication. The reporters
thus have exclusive scientific scoops, which allow them to select the top-
ics in advance, to contact the authors, and to investigate the subject before the scientific publication appears.[59] Nowadays, reporters learn the
latest scientific news before researchers do.

For the scientist, appearance in one of the most cited journals assures a certain notoriety, which can be greatly amplified if the press gets
hold of the story. This publicity network is a general rule, well recognized in scientific circles. The whole business is quite Machiavellian. To
achieve a degree of recognition among one's peers, the scientist at the
end of the twentieth century has to master the complex and subtle game
of scientific and media communication. This quest for the holy grail
of recognition can become an obsession and easily pervert the research
program of some scientists, who are willing to sacrifice everything to
achieve this consecration. Even if reporters do not systematically echo
all the published research—far from it—the key point is that the studies that are publicized have been selected by the major journals.

A SYSTEM DEMOLISHED

Whatever its imperfections and faults, the publication system has given
science its legitimacy and its universal dimension. But the system today

is seriously harmed by three recent tendencies in scientific communication. First, the rapprochement that has been achieved these last few years between publicly funded research and the industrial sector has a consequence: in many areas of research, patents replace or supersede primary publications. Moreover, the secrecy that surrounds some scientific results, objects of great commercial value, hampers the free diffusion of information. The second tendency is the inverse; research teams are beginning to disseminate their results directly by Internet, without peer review. Finally, at the end of the 1980s, we witnessed the appearance of a new form of communication, called "science by press conference" by the journal *Nature*—researchers choose to announce their discoveries to the press rather than through scientific journals to their colleagues, thereby transgressing one of the primary rules of the scientific community. The goal is to say loud and clear, "I am the first!" The prize is the attention of the government or industry with the apparent hope that major financial remuneration will follow.

The modes of circulation of scientific information have thus been profoundly modified.[60] The stakes are high. They concern the very mechanisms of funding research. Not long ago, research funding from the national government was granted to the universities and large research institutes; these were charged with distributing resources among diverse disciplines. But research has escaped neither the economic crisis that afflicts the industrial nations nor the worldwide tendency towards the Darwinian model of growth. To continue to work, research teams are constrained to find external funding from other institutions or from industry. It is apparent, in this context, that an article that appears in a scientific journal that is unknown to the public will have hardly any effect in attracting possible funding partners. The traditional fruit of research—the publication—is no longer the major criterion for the new funders of research; the financing of newsworthy science trumps that of ordinary science. This tendency has obvious consequences for the direction of science, because fundamental research, especially in the sciences of nature, has no short-term applications and unfortunately interests fewer and fewer politicians and managers.

Conscious of the stakes that the diffusion of research results represents, political leaders themselves encourage researchers to leave their laboratories and to communicate with the public.[61] For example, the "Environment, Research, and Society" days, organized in 1994 under the aegis of CNRS with the participation of all the large French research institutions, was an occasion to publicize research implications through the media: "It was conceded that the publication of research results in specialized journals is only one step in the communication process: it is an insufficient step. Other means of communication, real stages in the transfer of research results, are insufficiently developed."[62] Today, the use of the media constitutes one of the procedures of legitimizing research even in the heart of scientific institutions (though they do not always admit this). Communications departments, guided by the senior administration, mount publicity campaigns for research that might interest the public. For example, the publicity surrounding space or deep-sea exploration has contributed heavily towards justifying this very expensive research. Press releases are disseminated to the media to publicize the flashiest research of the moment. Anything that appears in the press is recorded and tallied. For six months in 1994, the CNRS in France faxed to all laboratories a summary of the daily media news in which the place of honor was given to the publicized "house" researchers. The summary never took the time to send the titles of the most brilliant scientific publications in each discipline. Some institutions have become proactive. American universities, often private and competing with one another, have their own publicity departments assigned to enhance the university image through media attention to research results.

Science has become a market controlled by publicity, and thus a power struggle. So it is increasingly frequent that research themes are oriented and financed as a function of their newsworthiness. In an era of tight financing, newsworthy subjects attract many research teams. Because directions validated by the news media must be pursued, the researcher's margin for maneuvering is reduced. Scientists' freedom of expression toward the public at large quickly reaches its limits when

they must consider the effect on the funders of research or the institutions that pay their salaries.

How can we reconcile the old system, based on professional scientific communication, the guarantor of real scientific progress, and the
new system, based on media communication, that increasingly deter-
mines the economic survival of laboratories? I will wager that colloquia
and public hearings will proliferate and lead to a new ethic, made necessary by the deregulation of scientific communication.[63]

THE FLAWS IN RESEARCH COMMUNICATION VIA THE MEDIA

Media hyping of research results is a two-edged sword. Whatever the
real scope of work or its degree of credibility, communicating it to the
public-at-large enhances the standing of its author. This development
establishes on the one hand the devaluation of traditional scientific research (scientific publications that have become too complex, inacces-
sible, disparaged), on the other hand the usefulness of communication
via the media. The latter is easy, quick, and confers status; benefiting
from a large audience, it reaches the decision-makers who fund research
and define priority objectives. But if the fallout from media hype can
benefit the researcher and his or her laboratory, paradoxically, one cannot devote oneself to the "questionable practices of stardom"[64] without
incurring the wrath of one's peers. For, despite official declarations, popularization is often seen in a bad light by the profession. A researcher
who grants an interview to a reporter transgresses the established order.
And if the reporter exaggerates a minor discovery or misinterprets a
story, the scientist takes the blame for having spoken inaccurately to an
interviewer who is often seen as an ignoramus. Worse, popular communication of scientific results still plays a minor role in the assessment of
scientists' work and in the development of their careers. Popularization
is not part of the worthy activities recognized by the scientific community. This is the reason why many prefer to remain in their ivory towers
and not to expose themselves to public view. This mistrust contrasts
with the striking expansion of the media and the strong demand of the

284   public. Those who agree to participate in the diffusion of scientific
knowledge do so because of taste, duty, personal or political interest, or
because they feel shielded from criticism by their positions.

In scientific communication by the media, four flaws should be deplored:

1. *Scientific information directly disseminated by the news media, because it is no longer vetted by one's scientific peers, is no longer reliable.* "The papal-type proclamation to the news media is overtaking the
critical, day-to-day work published in scientific journals. The age of
the prophets has returned and the truth is stated by those who speak
the loudest—or who have friends on editorial boards," wrote Alexandre
Ghazi, director of research at CNRS.[65] The risk is that the media echo
becomes the principal objective, beyond the value and precision of the
facts that are uncovered. This tendency is all the more disquieting because a news item spread by the media is usually judged as credible by
the public that sees it, whatever its foundation. This is the point correctly emphasized by the sociologist Jean Baudrillard in these terms:
"You disseminate a news item. So long as it is not falsified, it is convincing. And, except in the case of a lucky accident, it will never be falsified
in real time, so it will always be credible. Even if it is falsified, it will
never be absolutely false, since it was once credible. Unlike the truth,
credibility has no time limits; because it is virtual, it is never destroyed."[66] To falsify an untrue news item requires a lot of energy. The
result is always imperfect; doubt creeps in, belief persists, despite all the
efforts to convince and to reestablish the truth.

If the media are not vigilant, charlatans of all stripes, professional
sophists, prophets, defenders of esoteric pseudosciences, and mystics will
dominate the future.[67] The great media exposure of more or less serious
scientists nowadays, justified to varying extents, perturbs scientific debate. The question arises as to what constitutes a "scientific discovery."
Even when validated by a reputable journal, a discovery is not always a
discovery. To assess its value, one must follow the debates that it generates and wait until it is confirmed by other, independent teams. The
saga of "water memory" (a "discovery" published by the journal *Na-*

ture) is instructive in this regard. The many attempts to demonstrate the therapeutic effect of homeopathic medicines never won over the majority of medical researchers, who generally refused to grant the principle any value other than that of the placebo effect.[68] Conducting sound research to refute the concepts of water memory and homeopathic medicine is a long and exacting task, and its constraints are not easily compatible with the demands of the press: competition, the race against time, and the need to generate news, even if it is incomplete or imprecise.

2. *The quality of the media presentation is most important.* Information will be taken into consideration to the extent that its communication is associated with positions of great authority. Just holding an important position is enough to impress some reporters. Even an incompetent administrative director of a prestigious research institute will be queried by the press instead of a subordinate researcher, recognized only in his or her scientific field. The appeal to authority governs,[69] to the detriment of the quality of the information. Many exploit this advantage. The interviewee, to buttress his or her statements, emphasizes his or her titles, and the journalist makes note of them to reinforce to the editors the claim that the article is serious. The real value of prestige in the scientific hierarchy remains highly overestimated in some countries, and notably in France.

Other researchers will be solicited only because of their personal charm or media attractiveness. It is well known that a good researcher who expresses himself or herself poorly will not succeed. The teaching professors at universities, who devote a large part of their time to popularizing scientific knowledge, are greatly favored when it comes to popular communication in the media. It would be wrong to begrudge them this advantage.

3. *Not all scientific disciplines arouse media interest.* There exists a gradient of media interest in scientific information. In particular, two myths dominate popular articles.[70] One is the myth of the perfect world, which deals with the betterment of our life and our health. The media readily publicizes discoveries relevant to better protection of our envi-

ronment and new medical treatments. The other is the myth of the cosmogonic and eschatological dream, which incorporates studies on the origin or end of mankind: paleontology, astrophysics, environmental degradation, great catastrophes, the spread of great epidemics—all these fields fall in this category. By contrast, the media "penalize" basic scientific research, not easily explained outside a narrow circle of specialists. This problem tends to afflict research in mathematics, chemistry, physics, and some subdisciplines of biology.

Reporters' interest in science also depends on current events. To be picked up by the media, a piece of scientific information should, if possible, constitute an event. Research that constitutes a scoop is most frequently publicized. A corollary is that information in the media is ephemeral. An even more sensational happening dispels the previous one, which becomes banal and is quickly forgotten.

4. *Media polemics are not scientific controversies.* Controversy is an integral part of the history of science. Divergent hypotheses, contradictory theories, and challenged observations all stimulate scientific thought. With time, we generally succeed in distinguishing the true from the false. Controversy is formalized through the opposition of two contradictory arguments. Provisional recognition goes to the one that presents the best explanation or proposes an irrefutable demonstration. In principle, scientific controversy remains courteous and rests solely on scientific arguments. It can become violent and underhanded when the rhetoric is based on multiple baroque strategies aimed at convincing the widest circle of colleagues. The more intense the controversy, the more technical the evidence brought to bear on it. The form and the presentation of the arguments have to convince one's peers, which implies a specific language and citation of specialized publications. Sociologists have deciphered the customs and rules of these fratricidal jousts, which ought logically to lead to the establishment of the truth.[71] Once one viewpoint is conceded, it becomes everyday knowledge.

A polemic is of a different nature. Etymology clarifies the difference; the word comes from the Greek "polemikos," referring to a hostile or aggressive party in a war. The polemicist researcher, who often at-

tacks scientific theses or arguments in an extreme manner, manifests a
bellicose behavior. To win points, the polemicist would not hesitate to
point to motivations unrelated to the scientific subject. He or she would,
for example, accuse an opponent of seeking publicity in order to garner
research contracts. For want of substantive scientific arguments, the polemicist would not present arguments in a scientific publication; they
would be expressed only through the media. This behavior is often that
of someone who has lost a scientific argument but engages in a media
polemic in order to distract attention from the loss and to attempt to
convince administrators and decision-makers.

Among the various scientific polemics of the past decade, it is important to distinguish those hypotheses that were confirmed, those that
were disproved, and those that remain uncertain. Whatever the degree
of credibility of the polemicists, they have left a permanent mark on the
generation that witnessed the raising of the media stakes. Are the true
discoveries real? Is there not a grain of truth in the discoveries that have
been falsified?[72]

Some examples illustrate this conflict between truth and credibility.
Hibernatus, the "Ice Man" mummified by freezing and found in 1991
in a glacier at the border of Austria and Italy, really lived 5,000 years
ago. But how many people still believe that he is a hoax, as some illustrious scientists proclaimed several days after the announcement of his discovery? Near Marseilles, a diver, Henri Cosquer, discovered an underwater grotto decorated with extraordinary prehistoric engravings. Many
people, including recognized scientists, believed that this grotto was in
fact only a display of contemporary graffiti, the work of an artist who
tried to gain the lucrative exclusive rights to payments by the media.
They affirmed this without even having seen the drawings discovered
by local researchers. Although these illustrious "colleagues" have been
pilloried in a book devoted to this grotto,[73] here and there one encoun-
ters skeptics who still doubt the discovery. *Hibernatus* and the Cosquer
grotto are real discoveries, unfairly contested by scientific polemicists by
means of obliging or fooled media.

By contrast, water memory and cold fusion are, until shown to be

288　otherwise, unconfirmed "discoveries." Several research teams have attempted to duplicate the experiments but have been unable to confirm the heralded results. The anticipated or promised applications have never seen the light of day. The scientific controversy surrounding both discoveries was very heated, and polemics raged.[74] Doubt remains.

We should finally cite the greenhouse effect and the hole in the ozone layer. These are very real phenomena over which doubts still hover. What will be the scope of the global disturbances they induce? What will be the eventual consequences? Is it possible to act on the currently perceived causes? Opponents on these issues clash in specialized journals and in the popular media. The scientific uncertainty and the considerable underlying economic, social, and ecological stakes explain the conflicting information disseminated on these questions.

On the other hand, the problem posed by the introduction of the alga *Caulerpa taxifolia* never gave rise to a real controversy. From 1991 to 1995, several dozen publications of international rank and nearly a hundred communications in colloquium proceedings were published entirely by alarmist researchers; there was thus never a scientific debate in specialized journals. We had to wait until the end of 1995 for the defenders of *Caulerpa* to publish their hypotheses; but this attempt, as we have seen, was error ridden. On the other hand, over the same period, the affair of the "killer alga" was the object of an enormous media polemic that is barely beginning to wane. This case illustrates well the evolution of the communication of scientific information. To challenge scientists, it is no longer absolutely necessary to publish in specialized journals: by dividing opinion, a well-conducted media polemic can gain time and serve hidden interests.

The courteous, formulaic statement—"scientific opinions are divided"—often hides manipulations aimed at economic or political goals, especially if the subject is abstract or includes uncertainty that worries the public. Of course, the defense of interests other than those of basic research is not a recent phenomenon. It has often been attempted and sometimes has succeeded in corrupting scientific authorities.[75] Individuals who have taken this route are scattered throughout the long history of science.[76] But nowadays these anomalies are greatly amplified by the powerful influence of modern means of communication. If research institutions are expected to clean up their act, then journalists should be expected to separate the wheat from the chaff, reality from myth. It is often easier for journalists "without assuming responsibility for their remarks, to hype subjects, . . . to blow some point of view out of proportion, in short to generate considerable confusion and excitement, hiding behind the fact that the 'experts themselves are unable to agree,'" suggests Suren Erkman, himself a science reporter. "Instead of aiding the lay audience to form an enlightened opinion, the media very often only increase confusion."[77]

ON THE RESPONSIBILITY OF THE SCIENTIST

In the stories of *Hibernatus*, the Cosquer grotto, cold fusion, and water memory, the debate can be boiled down to just one question: is this a real discovery? In another debate, concerning the discovery of the virus causing AIDS, the only question was who discovered it, the French team of Luc Montagnier or the American team of Robert Gallo? In all these cases, the facts at issue were "static"; they have not changed. The controversies or polemics these cases aroused concern only scientific, economic, honorific, or legal stakes.

By contrast, in the contaminated-blood scandal in France, the mad cow disease crisis in England, the asbestos affair, global disturbances (greenhouse effect or ozone hole), and the *Caulerpa* problem, the underlying risk is "dynamic," for it concerns our future. While government experts and politicians fight over the assessment of the danger or the measures to adopt, the problem worsens. While the scientific controversy and the media polemic rage, the concentration of atmospheric carbon dioxide increases, the ozone hole grows. While meetings follow one another at a slow pace set by the overcommitted calendars of the experts, while the minutes are laboriously amended in richly paneled rooms, while supplementary assessments are awaited . . . the contaminated blood kills, the asbestos fibers suffocate, animal feed propagates

290   the prions, and the alga *Caulerpa taxifolia* spreads quietly. All with the perfect knowledge of the specialists.

Reversal is impossible; the dead cannot be resuscitated, an extinct species is gone forever, and it will take decades, indeed centuries, for certain destroyed habitats to return to an equilibrium with their former biodiversity. And the hopes placed in the development of new, sophisticated technologies will change nothing. The technical fixes that the prestigious signatories of the Heidelberg appeal place such confidence in represent an escape to the future. The sole concrete effect of this behavior is to delay the necessary decisions, while there is still time.

Faced with a dynamic risk, time alone counts. Everything possible should be done to intervene at the least warning and to prevent the debate from degenerating into vain polemics. When the world conference on the environment and development was held at Rio de Janeiro in 1992, the participants understood the time factor well in adopting the famous precautionary principle, a principle that every scientist, every politician, every expert should erect as a motto. In the affair of the "killer alga," the most deplorable aspect was that the little war among experts and scientists caused the loss of precious time while the situation spiraled out of control.

This affair has convinced me that, in such situations, scientific communication conducted according to the "rules of the game" is not effective.[78] In fact, what happens? Scientists discover a problem and evaluate the consequences. They communicate their results and analysis by means of the obligatory vectors of scientific and technical information, specialized journals. Scientists often stop there, content with having accomplished the normal functions of a researcher. Many forces keep them at this level: fear of crossing their hierarchical superiors, politicians, funders of research, not to mention concern for their careers, as the popularization of knowledge outside the institutional circuit is not valued—in fact, quite the contrary.

The history of scientific "affairs" at the end of this century is edifying. A bibliographic review of scientific communications shows that the

problems have often been very quickly posed correctly and the risks
properly assessed, but in specialized journals, usually incomprehensible
to decision-makers. Their assessments reach only one targeted audience,
the small community of research specialists on the subject, by definition
limited. In other respects, their impacts are almost nil: while a mobilization of the authorities with jurisdiction would be necessary, changing
research directions is difficult during a budgetary crisis; a reorientation
of research, even for the study of a major risk to our society, is too risky.
The traditional product of scientific work thus remains, in this sort of
situation, ineffective.

Some scientists are emboldened and warn the relevant authorities.
But the authorities do not like to hear from Cassandras, because it is
difficult for elected officials to enhance their standing by managing a
controversial problem, one that is likely to produce a tragedy. The chro-
nological repetition of aspects of the various incidents I have referred to
shows how many letters that were either misplaced or put in obscure
files either sank from sight or resurfaced only during legal examination
of a file. This fact bears witness to the mutual incomprehension between
scientists and administrators.

As for joint action with decision-makers and assessors, it is quickly
mired in the byzantine games surrounding the defense of various interests, careers, lobbies. It founders in the abyss of committee meetings
and closed-door sessions, where the only object is to demonstrate that
someone is paying attention to the problem—which assures the problem a discreet burial. There is no neutral university structure that allows
stakeholders to review the growing risks to our societies or the environment. Joint action faces obstacles that neither ethics committees nor
ethical rules will be able to clear.

Who is left to confront the problem? There remain the media. They
have abundantly proved their famed efficiency. They are the ones who
unmasked the transfusion of contaminated blood. They are also the ones
who shook the high-flying scientists and government experts from their
lethargy in the study of the effects of asbestos, by striving to apply the

alchemists' motto: *ad obscurius per obscura* (go towards the more obscure by way of the obscure). They are also the ones who warned the populations exposed to radiation in the Chernobyl fallout.

As I have said, to appeal to the media is to take a step that is often death for a research scientist's career, because this action is deemed so vulgar, self-interested, detestable, and narcissistic by the scientific establishment. However, if a scientist discovers a major threat and decides to go public, should we take as the correct model the one who is content to publish in specialized journals, the one who is moved to go further and to collaborate with government experts and politicians despite the inertia of the system, or the one who crosses the threshold and shouts his or her conviction in all directions to produce the only useful result: public concern with the problem?

We are responsible for safeguarding the natural environment for future generations. This is why scientific investigation of environmental damage is especially justified and necessary. But it also generates controversy and heated debates because the facts that observation and experiment establish are rarely neutral; they often describe damage caused by humans, thus raising the issue of responsibility. Then the socioeconomic, philosophical, and political considerations roused by conflicts of interest quickly overtake scientific considerations. This is the particularly difficult context in which ecological researchers are permanently stuck. Possessing knowledge, they have a moral duty to disseminate it beyond the small circle of readers of professional journals. Experts on nature, they have a social responsibility to inform society. The duty of dialogue should trump the right to remain silent or the hypocritical right to remain "prudent" that every bureaucrat claims. Scientists unable to make themselves heard in the chambers of authoritative decision-makers should turn to the media, the ultimate weapon of communication, to sensitize the decision-makers as quickly as possible and to constrain them to assume their responsibilities.

The public rightly complains of mistakes recognized too late. The scandals in which unscrupulous scientific administrators have played key roles repeat themselves. Ethical missteps that would be unthinkable in everyday life-threatening emergencies are effaced in the muffled cocoon of fearful professions, often more lax than protective. And the confidence of citizens towards politicians and scientists deteriorates. These reflections lead me to share the conclusion of Michel Massenet in his analysis of the French affair of blood contaminated by the AIDS virus: "Every democracy knows scandal: individual and collective misdeeds that affect society, the institutional dysfunctions, the perverse games of power and money do not remain hidden, and we should rejoice in this fact. The remedy to these problems is found in the freedom and audacity of the media, and especially of the press." [79]

We should recognize the salubrious role of the press in the *Caulerpa* affair: they told the story of a threat to our environment. They contributed to forcing attention on this affair at a time when diverse powers and local interests wanted to bury it. Of course, we can regret the deplorable exploitation of the same media lever by scientists who expressed themselves for several years only through reporters. Of course, we can deplore the excesses of some reporters, how they blew a false scientific controversy out of all proportion in order to attract more readers. But the media constitute a locus for free expression in democratic, liberal nations that scientists should know how to use advisedly. Researchers ought to have, if not the obligation, at least the responsibility to turn to the press if they discover a major threat and their warnings, for various reasons, are being stifled. The progression of scientific discourse in the media is inescapable; we must accept it. It is incumbent on scientific and journalist partners to publicize the perversion of elementary scientific principles of morality and democracy. The culture of science—its critical analyses, codes of conduct, disciplinary rules, and ethics—are useful tools in framing a new method of disseminating scientific information while maintaining the essential freedom of expression. Scientific communication via the media should remain readily available and allow for the expression of matters of conscience.

May these three "lessons of *Caulerpa*," and the story of my initial attempt to take account of a major environmental threat, help to enrich reflection on the human role in the management of our natural heritage.

The Biology of *Caulerpa* taxifolia as Known in 1991[1]

Anatomy and Morphology of the Beautiful Stranger

The algae belonging to the group Caulerpales have a specific anatomy: they have no internal cell walls and therefore no specialized cells. When they are cut, a sap flows out while a scarlike, fibrous plug, composed primarily of carbohydrates, quickly forms and prevents further loss of sap. This liquid, which occupies the cavity running from the tips of the "leaves" to the ends of the "roots," contains the organelles, including nuclei, in great profusion. The entire alga is thus a length of pipe, a single cell: it is termed siphonalean or coenocytic. Individuals of *Caulerpa* have been found that are 2.8 meters long: these are the largest known single cells.[2] Only several groups of green algae and fungi have this peculiar structure. The rigidity of the alga comes from the tens of thousands of membranous pillars ("trabeculae") that go from one side of the siphon wall to the other, like struts.

Our observations, recorded during dives beginning in the summer of 1990, allowed us to describe the morphological adaptations of the alga in its new Mediterranean habitat. The alga comprises three easily identified elements: the stolon, the fronds, and the rhizoids.

The main part is a horizontal axis that creeps along the bottom: this is the stolon, similar to that of a strawberry plant. It can reach over two meters in length and is one to two millimeters in diameter. Depending on the type of bottom and density of vegetation, it is more or less branched. Stolons are easily observed at the edges of colonized zones: a multitude of axial tips seem to be creeping outward in search of virgin territory. This peculiarity must have struck the French naturalist Jean Vincent Félix Lamouroux who, in 1802, was first to describe the genus *Caulerpa* and chose this name by joining two Greek words, *caulos* ("stem") and *erpo* ("to creep").

The axes bear "leaves" towards their upper surface. This term is incorrect in the specialized vocabulary of phycologists (it is reserved for more complexly structured plants); we prefer "frond." The fronds are strikingly beautiful, as much for their brilliant, fluorescent green color as for their finely dissected, feathery form (each frond carries on various parts of its axis several dozen pinnules). An axis can carry more than a hundred of them. Their size is highly variable; they tend to be short (five to ten centimeters) wherever there is much light, thus at shallow depths; by contrast, when the plant receives little light (in late autumn, in winter, at substantial depths, in the shadow of rocks, etc.), their length can surpass sixty centimeters!

The axes run over marine soils, fastening themselves with a type of root, a term that is again incorrect in phycological parlance: among algae, these structures are primitive and they are called "rhizoids." These penetrate every surface and are adapted to the texture of the bottom. They are long in sand or mud and short when they are insinuated in tiny rock crevices and pits. When the plant colonizes *Posidonia* meadows, the rhizoids can even form clusters of stilts like red mangrove trees: these elevate the axes and permit the *Caulerpa* fronds to overtop the *Posidonia* leaves. Certain species of *Caulerpa* can assimilate organic substances: this ability is unknown among other algae and green plants, but it is typical of fungi.

Stolon, fronds, and rhizoids form the vegetative apparatus of the plant, commonly called the "thallus." Its architecture is surprising: it resembles that of the higher plants, but it is unicellular. Phycologists have asked how the genetic information in the nuclei leads to the formation of such an elaborate architecture without the aid of the classical accumulation of "building blocks" (the cells) for which the form of assembly is programmed by the order of formation of different cells.

The thallus (the alga) grows especially rapidly when the water temperature exceeds 18° C (from June to November). The axes elongate at that time by almost two centimeters daily and can form a new frond and a bundle of rhizoids every two days. While the axis elongates at one end, it dies at the other, so that no part of the thallus lives more than a year. Thus, it is always the same individual, which perpetuates itself by unlimited growth.

This entire structure leads rapidly to dense colonies, so dense that they compete with native plant species and sessile animals. In several years, the entire invaded area is covered and the great majority of Mediterranean species disappear under the thick lawn of *Caulerpa*. Quadrats of vegetation have yielded up to 8,200 fronds per square meter! Placed end to end, that often represents more than two kilometers of fronds borne on more than 244 meters of axes per square meter! Marking of well-mapped circular colonies has allowed us to estimate an annual extension of the diameter in the range of three meters. The species of algae and sessile invertebrates that can survive under this sort of carpet are rare indeed.

In winter and spring, the alga does not disappear, but the ends of the fronds become blanched and wither. They branch in all directions; thus, a frond can bear several dozen secondary fronds. The axes die at various spots. The individual becomes fragmented. During the spring storms, many *Caulerpa taxifolia* fragments are torn up. Now, each fragment of any part of the alga, so long as it contains a nucleus, can initiate a new alga; almost all fragments contain many nuclei. A frond fragment one centimeter long, attached to the bottom in early spring, can produce a dense colony three meters in diameter by late autumn. This mode of vegetative reproduction by natural breaking is therefore extremely effective and works well in the Mediterranean.

Everything for Sexual Reproduction

To reproduce sexually, algae in the Caulerpales have an unusual strategy: several days each year, in certain individuals, each nucleus joins with a chloroplast (the cell organelle containing chlorophyll), surrounds itself with a membrane, and is thus transformed into a male or female gamete. These gametes are all expelled into the water through small orifices, which causes the entire plant to die because it is completely emptied of its contents; the only thing that is left is the wall of the single giant cell. A *Caulerpa* individual that has just reproduced sexually becomes a brilliant white before fragmenting and dissolving.

Caulerpa taxifolia is monoecious (the two sexes are borne by the same individual plant); near the base of the frond are found the male gametes; toward the tips of the frond are the slightly larger female gametes. The "sperm" and the "ovum" of *Caulerpa taxifolia* barely reach five microns (thousandths of a millimeter). They are very similar to one another and they are shaped like a drop of water with two little hairs (the flagella) that allow them to move. Each gamete has one nucleus plus one chloroplast.

The fusion of a male and a female gamete produces an egg (zygote) that grows through two intermediate stages before becoming an adult plant. The life cycle thus appears to comprise a single generation: an individual reproduces and creates a new individual which reproduces sexually in the same manner. The life cycle of *Caulerpa*, described in detail by two Japanese specialists, indicates that the passage from the egg stage to the adult plant takes three to six months. How the nuclei divide in the different stages of growth remains unknown. The egg always has two times as many chromosomes as the gametes, but we do not know the developmental stage during which the genetic contents of the nuclei are halved (the process of meiosis).

Sexual reproduction has never been observed in the clone of *Caulerpa taxifolia* introduced to the Mediterranean. The existence of a colonization site in Toulon (150 kilometers from the location of the initial colony, Monaco) led us to fear that eggs were being carried very far by

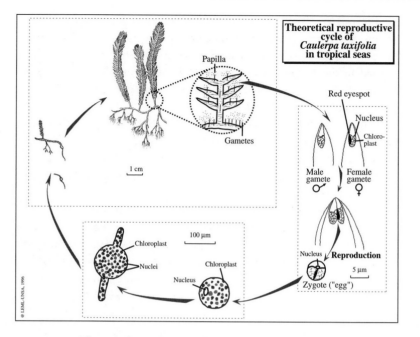

FIGURE A1.1.   Theoretical reproductive cycle of *Caulerpa taxifolia* in tropical seas

prevailing currents or on boat anchors and that they might be able to establish themselves in favorable sites. However, subsequent research confirmed that all gametes in the Mediterranean are male.

Its Ecological Niche and Resistance to Winter Temperatures

The vicinity of Cap Martin (located five kilometers east of Monaco) contains, unlike the cape of Monaco, very diverse substrates (rock, sand, mud) with a rich natural vegetation. Several hundred algal species and three marine flowering plants (phanerogams) can be seen there. In the autumn of 1990, *Caulerpa taxifolia* began here and there to penetrate and to overgrow the meadows of the largest marine phanerogam of the

Mediterranean: *Posidonia oceanica*. This is far from an anecdotal threat because the regeneration of a *Posidonia* meadow takes a very long time: this plant requires a century to progress three meters! This characteristic has led to the frequent comparison of *Posidonia* meadows to tropical forests: they occupy an analogous role for coastal marine ecosystems, and they are similarly fragile.

The *Caulerpa taxifolia* clone introduced to the Mediterranean has a peculiar trait: its resistance to the cold winter waters. Thus, we were able to establish by 1990 that the invasion of this tropical species is completely independent of global warming. Its cold resistance must be attributable to remarkable genetic characteristics, different from those of its "cousins," the *Caulerpa taxifolia* that typically inhabit all the tropical seas. These latter individuals have never ventured to the coasts of adjacent temperate seas: they die when the water temperature falls below 20° C!

On our shores, the temperature of the coldest seawater (below 13° C) is very harmful to the common native Mediterranean *Caulerpa*, *Caulerpa prolifera*, a species confined on the continental French Mediterranean coast to some twenty favorable sites. We had, from the outset, observed that *Caulerpa taxifolia* spread to areas where *Caulerpa prolifera* would never have been able to grow. In November 1990, in the Gulf of Juan (forty kilometers west of Monaco), the drop in water temperature caused necrosis of the apical extremity in every frond of *Caulerpa prolifera*, while on the same date no *Caulerpa taxifolia* frond showed any such damage. It was not until the end of December that nearly ten percent of *Caulerpa taxifolia* fronds near the surface displayed this sort of necrosis. The surface seawater temperature is regularly recorded in front of the Oceanographic Museum of Monaco. Between 1984 and 1990, there were 56 days when the water temperature dropped below 12.5° C, a cold temperature for that part of the Mediterranean. However, in the winter, *Caulerpa taxifolia* proves very resistant: in water five meters deep, where the temperature drops below 13° C in the winter, the extremities of the fronds are blanched, the visible stolons show many necroses, the alga becomes etiolated but does not die. In the summer, it begins to grow as soon as the water temperature surpasses 18° C. The

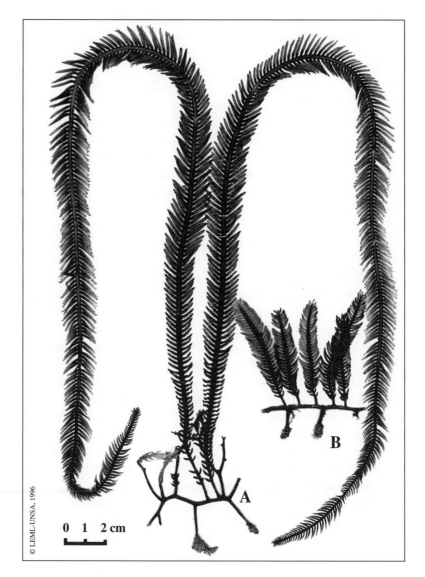

FIGURE A1.2. Phenotypes of *Caulerpa taxifolia* in different environmental conditions

speed at which the alga spreads depends on the number of days when the water temperature is above this threshold.

The Genus Caulerpa

The genus *Caulerpa* includes nearly a hundred species and varieties of very diverse shapes. They are distinguished essentially by morphological criteria. In certain groups, the form varies greatly depending on environmental conditions, which makes it difficult to recognize the species. But for the majority of species of *Caulerpa*, this morphological plasticity is not extreme and their identification is quite easy.

Caulerpa taxifolia was first discovered in the tropical Atlantic Ocean (in the vicinity of the Virgin Islands). In this ocean, it was found from the Caribbean to Brazil and in the east on the western tropical coasts of Africa. It is also known from the Indian Ocean on the African coast and from Pakistan to Indonesia. In the Pacific Ocean, it is known from Japan to Australia. Although it was recently recorded in the Red Sea, it had never been described in the entire Mediterranean.

In the Mediterranean, five other *Caulerpa* species are present:

1. The most common species is doubtless *Caulerpa prolifera*. Unlike *Caulerpa taxifolia*, *Caulerpa prolifera* has always been in the Mediterranean, but on the continental French coast, it was known only from about twenty sites in which it was well established. In these areas it most frequently grows in small bottom patches (between 0 and 10 meters deep) in calm water on a sand-mud soil or in a mat of dead *Posidonia*. The ecological niche of this *Caulerpa*, with flattened, toothless fronds, is thus much narrower than that of *Caulerpa taxifolia*.

2. *Caulerpa ollivieri* is a rare species, the validity of which is debatable (it may simply be a dwarf form of *Caulerpa prolifera*).

3. *Caulerpa racemosa* has fronds resembling tiny bunches of grapes. In 1990, it was known only near the coasts of Tunisia, Egypt, Lebanon, and Syria. Since then, this small *Caulerpa* has been observed in Greece and Italy (in Sicily and near Leghorn). It thus appears to be progressing

towards the northern coasts of the Mediterranean.

4. *Caulerpa mexicana,* with flattened, pinnate fronds, but in which the pinnules are always wider and shorter than those of *Caulerpa taxifolia,* has been recorded only from the coasts of Israel and Syria, under the name of *Caulerpa crassifolia.* This species is rare; since its discovery in 1941, it has been found by only four researchers.

5. *Caulerpa scalpelliformis* also has feathery leaves with very wide pinnules. It has been found only off the coasts of Israel, Lebanon, and Syria.

The presence of *Caulerpa prolifera* in the entire Mediterranean and of *Caulerpa racemosa, Caulerpa mexicana,* and *Caulerpa scalpelliformis* in the southern Mediterranean has often been discussed. To explain the origin of these common elements of the tropical marine flora, people have been torn between hypothesizing a recent lessepsian origin or that these are relics of an ancient tropical Mediterranean Sea, originally derived from the Atlantic Ocean.

In other seas, there have never been recorded proliferations of *Caulerpa taxifolia* such as those we encountered at Monaco and Cap Martin. Records of frond size found in bibliographic references give a mean length between two and twenty centimeters, similar to that we measured in our herbarium samples from tropical regions. The sizes of the fronds (up to eighty-five centimeters) seen at Monaco and Cap Martin, in very dense populations and exposed to little light, are thus exceptional for this species.

In February 1991, the sum of these preliminary data led us to give a conservative opinion on its possible eventual eradication. The discovery of the alga at Toulon led us to suspect the existence of many other contaminated sites between Monaco and Toulon. As we then suspected great powers of dissemination and did not yet know the role of sexual reproduction, we were skeptical about the chances for success of a battle against the alga. Thus, we wrote then: "The present spread of this species over several hectares along a minimum of 150 kilometers of coast, and its very efficient reproductive strategy, combine to make any manual eradication unlikely."

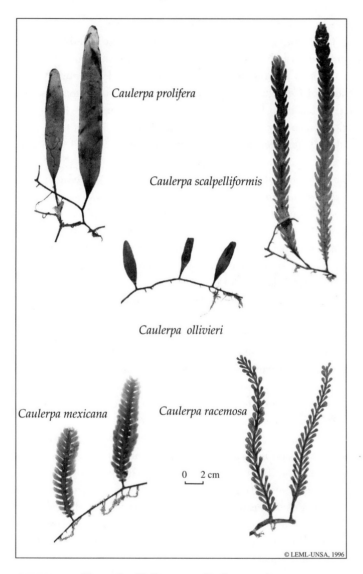

Caulerpa prolifera

Caulerpa scalpelliformis

Caulerpa ollivieri

Caulerpa mexicana   *Caulerpa racemosa*

0   2 cm

© LEML-UNSA, 1996

FIGURE A1.3.   Five native Mediterranean *Caulerpa* species

Chronology of a Heralded Invasion

1980 • A clone of *Caulerpa taxifolia* resistant to the cold is observed in the tropical aquarium at Stuttgart (Germany). It is distributed to the aquaria of Nancy, the Oceanographic Museum of Monaco, and the Oceanographic Institute of Paris.

1984 • A square meter of *Caulerpa taxifolia* grows in the Mediterranean beneath the Oceanographic Museum of Monaco.

1989 • The acclimation of the alga is discovered by the author (A. Meinesz). The first concerns are expressed verbally to the Monacan and French authorities.
 • A hectare of *Caulerpa taxifolia* is growing in front of the Oceanographic Museum of Monaco.

1990 • JULY—The alga is discovered in France at Cap Martin, five kilometers east of Monaco. First newspaper article, first letters of

concern to the French authorities (prefecture, Maritime
Affairs, general council). In the press, the directorate of the
Oceanographic Museum minimizes the threat.

- SEPTEMBER—The alga is found near Toulon, 150 kilometers
west of Monaco. Off Monaco and at Cap Martin, the alga covers three hectares.

1991
- JANUARY—The fishermen of Alpes-Maritimes are worried;
they drag up *Caulerpa* in their nets set more than 100 meters
deep.
- FEBRUARY—The first scientific publication recounting all the
risks associated with the invasion of the introduced alga is accepted in the journal *Oceanologica Acta.*
- APRIL—A file listing studies to be urgently undertaken is submitted to the Ministry of the Environment and the Secretariat
of State for the Sea. The director of the Museum turns to the
press to deny the Monacan origin of the alga and to proclaim
its benefits.
- MAY—First contact with IFREMER, resulting in no action.
- JULY—IFREMER and the directorate of the Oceanographic
Museum downplay the problem.
- AUTUMN—The alga spreads at Monaco and Cap Martin (thirty
hectares). It is found at Lavandou and Agay (Var).
- DECEMBER—It is found at Saint-Cyprien (Pyrénées-
Orientales), 350 kilometers west of Monaco. Scientists from
the Universities of Nice, Marseilles, Pisa, and Trent organize
a round table. They call for the eradication of the alga.

1992
- JANUARY—The alga is found at Saint-Cyr-les-Lecques (Var),
200 kilometers west of Monaco.
- FEBRUARY—Publication on the toxins contained in the alga.
- FEBRUARY AND OCTOBER—Four commissions are created, and

four ministers come to the Côte d'Azur. But no decision is   **307**
made to control the algal spread.

- JUNE—The alga is found at Imperia (Italy), forty kilometers east of Monaco.
- SEPTEMBER—The alga is found in the Balearic Islands (Spain).
- DECEMBER—European Union financing of the first research: 125 researchers in 25 laboratories can finally study the diverse aspects of the problem.
- LATE 1992—The alga spreads to twenty-eight sites between Menton and Toulon. Four hundred thirty hectares are more or less infested.

1993
- The alga spreads in Italy (Sicily, Elba, Leghorn, and in front of five sites in Liguria).

1994
- The alga is found in the Port-Cros National Park, one hundred kilometers west of Monaco.
- DECEMBER—The scientists assigned to study the problem issue the "Barcelona Declaration": the spread of *Caulerpa taxifolia* constitutes a major threat to Mediterranean ecosystems.
- LATE 1994—By this time, a total of thirty-eight sites have been invaded, and 1,500 hectares are infested. The alga is found in two sites in Croatia (Adriatic Sea), more than 1,000 kilometers east of Monaco.

1995
- LATE 1995—Monacan researchers publish an article in a journal of the Academy of Sciences trying to prove that the alga came from the Red Sea and is innocuous.

1996
- JULY—Counter-article appears in the Academy of Sciences journal, arguing that certain results of the Monacan researchers are incorrect.

- SEPTEMBER—The European Community finances a second research program.
- LATE 1996—By this time, sixty-eight sites have been invaded, and 3,000 hectares infested.

1997
- MARCH—Workshop of the French Academy of Sciences with fifty specialists. The incorrect assertions of the Monacan article were not criticized. Ambiguous conclusions about the origin of and danger presented by the alga, but strong recommendation to attempt to prevent dissemination and to control the spread.
- MARCH—First edition of this book.
- SEPTEMBER—Third International Workshop on *Caulerpa taxifolia* at Marseilles. Researchers ask the prime minister to establish a strategy for prevention and control.

1998
- MARCH—United Nations Workshop for the Mediterranean countries at Candia. Representatives unanimously conclude that the alga constitutes a major threat to Mediterranean ecosystems and recommend that all countries establish a strategy to prevent dissemination and to control the invasion.
- OCTOBER—First study of *Caulerpa taxifolia* genetics. All aquarium strains and colonies from the Mediterranean are the same.
- DECEMBER—U.S. Department of Agriculture proposes adding aquarium clone of *Caulerpa taxifolia* to Federal noxious weed list.

1999
- JANUARY—Motion introduced to French Chamber of Deputies calling for inquiry into the polemic surrounding *Caulerpa taxifolia* and the slow response of the national government to the spread of the alga.
- APRIL—Aquarium clone of *Caulerpa taxifolia* added to U.S. Federal noxious weed list.

Table of Measures

1 millimeter = 0.039 inches

1 centimeter = 0.39 inches

1 meter = 3.28 feet

1 kilometer = 0.62 miles

1 hectare = 2.47 acres

1 kilogram = 2.20 pounds

10° centigrade = 50° Fahrenheit

20° centigrade = 68° Fahrenheit

1 French franc = $0.20 U.S., approximately

Acronyms of Organizations

AFP: Agence France-Presse

AMPN: Association monégasque de protection de la nature (Monacan Association for the Protection of Nature)

CCNE: Comité consultatif national pour l'éthique (National Advisory Committee on Ethics)

CEA: Commissariat à l'énergie atomique (Atomic Energy Commission)

CEVA: Centre d'études et de valorisation des algues (Center for Algal Research and Economic Development)

CIESM: Commission internationale pour l'exploration scientifique de la Méditerranée (International Commission for the Scientific Exploration of the Mediterranean)

CNRS: Centre national de la recherche scientifique (National Center for Scientific Research)

DIREN: Direction régionale à l'environnement (Regional Directorate of the Environment)

DRAE: Direction régionale à l'architecture et l'environnement (Regional Directorate of Architecture and the Environment)

ENEA: Comité national de recherche et de développement de l'énergie nucléaire et des énergies alternatives (National Committee for Research and Development of Nuclear and Alternative Energy)

312   FFESSM: Fédération française d'études et de sports sous-marins
(French Federation of Underwater Studies and Sports)

GISMER: Groupement d'intervention sous la mer (Undersea Intervention Team)

GIS-Posidonie: Groupement d'Intérêt Scientifique Posidonie (Scientific Working Group on Posidonia)

HTE: High Tech Environnement

ICRAM: Istituto centrale per la ricerca scientifica e technologica applicata alla pesca maritima (Central Institute [Italian] for Scientific and Technological Research on Marine Fisheries)

IFREMER: Institut français de recherches pour l'exploitation de la mer (French Research Institute for Marine Development)

INSERM: Institut national pour la santé et la recherche médicale (National Institute of Health and Medical Research)

NGO: non-governmental organization

ORSTOM: Office de Recherche Scientifique pour les Territoires d'Outre Mer (Office of Scientific Research for Overseas Territories)

POS: plan d'occupation des sols (master zoning plan)

RPR: Rassemblement pour la République (Rally for the Republic)

SRETIE: Service de la recherche, des études et du traitement de l'information sur l'environnement (Environmental Research and Data-Processing Service)

UCA: Union des conservateurs d'aquariums publics de France (Union of Curators of Public Aquaria in France)

UICN: Union internationale pour la conservation de la nature (International Union for the Conservation of Nature)

URVN: Union Régionale de la protection de la Vie, de la Nature et de l'Environnement (Regional Association for the Protection of Life, Nature, and the Environment)

ZNIEFF: zone naturelle d'intérêt écologique, floristique ou faunistique (natural area of ecological, botanical, or zoological concern)

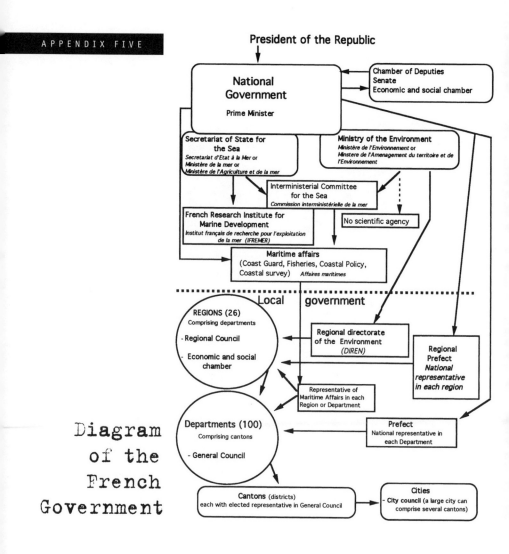

President of the Republic

National Government

Prime Minister

Chamber of Deputies
Senate
Economic and social chamber

Secretariat of State for the Sea
Secretariat d'Etat à la Mer or
Ministère de la mer or
Ministère de l'Agriculture et de la mer

Ministry of the Environment
Ministère de l'Environnement or
Ministere de l'Amenagement du territoire et de l'Environnement

Interministerial Committee for the Sea
Commission interministérielle de la mer

French Research Institute for Marine Development
Institut français de recherche pour l'exploitation de la mer (IFREMER)

No scientific agency

Maritime affairs
(Coast Guard, Fisheries, Coastal Policy, Coastal survey)   Affaires maritimes

Local   government

REGIONS (26)
Comprising departments
- Regional Council
- Economic and social chamber

Regional directorate of the Environment
(DIREN)

Regional Prefect
National representative in each region

Representative of Maritime Affairs in each Region or Department

Departments (100)
Comprising cantons
- General Council

Prefect
National representative in each Department

Diagram of the French Government

Cantons (districts)
each with elected representative in General Council

Cities
- City council (a large city can comprise several cantons)

Notes

FOREWORD

1. Charles S. Elton, *The Ecology of Invasions by Animals and Plants* (London: Metheun, 1958), 19.

2. Ibid., 18.

3. D. Simberloff, "The biology of invasions," in D. Simberloff, D.C. Schmitz, and T. C. Brown, eds., *Strangers in Paradise: The Impact and Management of Nonindigenous Species in Florida* (Washington, D.C.: Island Press, 1997), 3–17 at 3.

PREFACE

1. "Man and the Sea," in Charles Baudelaire, *Selected Poems from "Les Fleurs du Mal": A Bilingual Edition,* trans. Norman R. Shapiro (Chicago: University of Chicago Press, 1998), 25.

CHAPTER ONE

1. When the student came to talk to me about the introduction of the tropical *Caulerpa* into the Mediterranean, I had a total of twenty publications on this group of algae.

2. During my national service, I had been his midshipman in the Navy (his cadet), then his ship's ensign; he was then my captain.

3. The first building of the Oceanographic Institute, the foundation of which was set in 1906, was built on a site at the Sorbonne, rue Saint-Jacques in Paris. The Oceanographic Museum of Monaco, built later, was then given to this foundation.

4. The characteristics of the alga *Caulerpa taxifolia* in the Mediterranean resemble those of many invasive species that harm the environment, discussed in detail in chapter 7 and deplored in most ecology texts. A good definition is that of Jean Dorst, former administrator of the Foundation for the Oceanographic Institute, in *Before Nature Dies* ([Pelican, 1970], 218): "In their original milieu animals and plants occupy clearly determined ecological niches, where their populations are controlled by competition and predation. In a new biological association they may either disappear rapidly, being 'smothered' by the environment, or else they may become pests. In the long run their explosive success becomes catastrophic for natural habitats, native plants and animals, and often for the human economy."

5. The winters of 1985 and 1986 had been very cold; it had snowed heavily on the coast.

6. This species, common in the Mediterranean, is present on the Côte d'Azur, but it is found only in a few sites where conditions are favorable.

7. The large marine flowering plants, which are not algae, constitute the dominant vegetation on all Mediterranean coasts. Like the iris, they have rhizomes and leaves shaped like lashes more than a meter high, grouped into bundles. Very dense (sometimes more than 800 per square meter), these bundles of ribbon-like leaves constitute a meadow-like vegetation called a "grass bed," which is the equivalent of a terrestrial forest. Many marine animals feed on them or shelter in them. These are "keystone species" of coastal Mediterranean ecosystems.

8. The Ligurian current flows from east to west.

9. In fifteen years of close collaboration with the journal *Science et Vie*, this was the only subject on which her editor-in-chief turned her down. She could not conduct this inquiry until three years later, after a change in editor-in-chief (see her article, "The alga that poisons scientists," *Science et Vie*, September 1993). In May 1991, another reporter, Marguerite Tiberti, conducted a detailed inquest for the same journal. Her article was to be illustrated by slides of mine that she had selected and that I had sent to the editors. Her

article never appeared. She could not use her work on this subject until the directorate changed: an inset on *Caulerpa* was published in 1993 in a more general article on introduced species in the Mediterranean (*Science et Vie*, no. 907, April 1993).

10. Coincidence: the genus *Caulerpa* was described first in 1802 by the French naturalist Jean Lamouroux (1776–1825).

11. Here is what we knew about the biology of the genus *Caulerpa* in 1990 and presented in the report to the local authorities, very briefly summarized:

This algal genus is known to reproduce in two ways. One is by vegetative reproduction (natural suckering): fragments of "leaves" (fronds) or of stolons (the horizontal axes) can be ripped up by water movements, planted far away, and constitute new foci of colonization. The other is by sexual reproduction; in this case, all the nuclei of the siphonaceous alga (the thallus) are transformed into gametes. After they are expelled into the sea, gametes unite to produce an egg (zygote) that is capable of producing a new plant. At the time of the report (July 1990), I had not encountered in the Mediterranean any *Caulerpa taxifolia* showing the normal signs of sexual reproduction (which occurs in any event only a few days each year); all the isolated thalli arose vegetatively. One characteristic of the genus *Caulerpa* is that it contains two kinds of toxin, caulerpine and caulerpicine, which can accumulate in the flesh of fish that eat the alga and in turn cause discomfort for whoever eats these fish. Fish of the southern Mediterranean that frequently graze the common Mediterranean *Caulerpa* (*Caulerpa prolifera*) are also reported to be unfit for human consumption.

12. IFREMER = Institut français de recherches pour l'exploitation de la Mer (French Research Institute for Marine Development).

13. By contrast, in the northern Pacific and the Indian Ocean, other species of *Caulerpa*—from the Sedoideae section—are eaten and even cultivated in the Philippines Archipelago (*Caulerpa lentillifera* and *Caulerpa racemosa*). Thus the different species of *Caulerpa* can be distinguished by their edible qualities. This is a frequent phenomenon in nature; though we like the royal agaric mushroom, or Caesar's *Amanita*, other species of *Amanita* are often toxic, even lethal.

14. In fact, the IFREMER aquaculture laboratory confused *Caulerpa taxifolia* with southeast Asian *Caulerpa* species that are edible.

15. This association wanted to know the potential impact of a construction project of a harbor for pleasure-craft, which threatened to wreak great havoc on the marine ecosystems within its perimeter.

16. *Oceanologica Acta* 14, no. 4 (1991): 415–26.

17. These species are termed "lessepsian," after Ferdinand de Lesseps, French diplomat who had dared to conceive and then to achieve the construction of the Suez Canal, inaugurated in 1869.

18. "Phycologist," from the union of the Greek terms "phycos" (= alga) and "logos" (= study). This term is preferred to "algologist," either to avoid confusion with other scientists who justifiably lay claim to the title "algologues" (those who study pain [the suffix "algia," from the Greek "algos," pain]) or because it is improper to join the Latin "alga" and the Greek "logos."

19. "Informed by the divers of the Oceanographic Museum of Monaco of the proliferation of a tropical *Caulerpa* beneath the Museum, we dove in August 1989. In front of this coast, which is very exposed to wave action, the alga constituted a very dense, continuous cover at depths between five and thirty-five meters on all kinds of substrates (rock, sand, mud), competing with the native algal flora. We identified this alga as *Caulerpa taxifolia*, a species with a wide, pantropical distribution.

According to the testimony of an aquarium employee, the alga, which has been cultivated in several Museum tanks for about fifteen years, would have been observed beneath the Museum for the first time in 1984. Progressively, it would have colonized the small bottom patches located in front of the Museum essentially between five and thirty-five meters deep with several isolated stolons to forty meters" (*Oceanologica Acta*, 14, no. 4 [1991]: 417).

CHAPTER TWO

1. DRAE, now called DIREN = Direction régionale à l'environnement = Regional Directorate of the Environment.

2. SRETIE = Service de la recherche, des études et du traitement de l'information sur l'environnement = Environmental Research and Data-Processing Service.

3. University of Aix–Marseilles II.

4. At a general assembly of the local committee of marine fisheries of Alpes-Maritimes, Jean-Paul Roux, the elected prud'homme representing the professional fishermen of the Villefranche–Cap Ferrat–Beaulieu sector, asked Maritime Affairs to examine the problem (see *Nice-Matin*, February 18, 1991). The "prud'homme" is elected as the quasi-official spokesperson and represen-

tative of the fishermen of a region ("prud'homie"), interacting with government and agency officials on their behalf.

5. My responsibilities on many scientific committees for marine parks and reserves and those as president of the environmental commission of the Economic and Social Council of the region Provence–Alpes–Côte d'Azur had led me, six months previously, to write to the Minister of the Environment at the time, Brice Lalonde. I had wanted to draw his attention to the plight of field research (ecology, botany, zoology, geology) in our parks and reserves, which became increasingly more dire.

6. At a total cost of 300,000 francs (to be shared among the three associated laboratories).

7. For a total sum of 120,000 francs.

8. Soon after his arrival at the Museum, Doumenge even stated "that he found the institution a little old-fashioned, having hardly advanced since 1930!" (*Nice-Matin,* May 9, 1989).

9. European patent number 0328474 (published as number 0328474 B1, April 8, 1992), with French priority (submitted February 6, 1989), entitled "Process of biological purification of waters containing organic matter and products derived therefrom, using diffusion and the activity of aerobic and anaerobic microorganisms, and plan for its deployment." Patent-holder and inventor: Jean Jaubert.

10. The quantity moved was either 48 tons (according to *Science et Vie* of February 1990) or 4.8 tons (according to *Science et Vie* of August 1990).

11. *Science et Vie,* February and August 1990; *Nice-Matin,* October 7, 1989, November 18, 1989, and August 26, 1990.

12. Research structure enacted under the aegis of the "open partial accord on major natural and technological risks" of the Council of Europe.

13. *Nice-Matin,* February 10, 1993.

14. On the internet page of the Enoshima aquarium (Japan) can be found a presentation on Micro-ocean by the Oceanographic Museum of Monaco (http://www.iseshima.com/enoshima/monaco.html.)

15. *Science et Vie,* February 1990. See also *Nice-Matin* of October 7, 1989: "'To save the coral from the pillage and devastation to which it is increasingly subjected by collectors and fishermen is a task as urgent as saving the Amazonian forests,' emphasized Professor Doumenge, director of the Oceanographic Museum. 'To be able to have captive propagation of corals is thus of major interest.'"

16. See *l'Événement du Jeudi*, January 31, 1991; *Le Marin*, February 22, 1991; *Science et Vie*, May 1991; *Agence France-Presse*, dispatch of July 1, 1991; *Nice-Matin*, July 2, 1991; *Reuters*, dispatch of July 14, 1991.

17. *Agence France-Presse*, dispatch of July 1, 1991; *Nice-Matin*, July 2, 1991; *Reuters*, dispatch of July 14, 1991.

18. *Nice-Matin*, October 7, 1989.

19. *Reuters*, dispatch of July 14, 1991. In fact, at this time, it was generally conceded that corals could fix only a tenth of the CO_2 that tropical forests could. A year later, it was determined that coral reef ecosystems are, on the contrary, sources of atmospheric CO_2. (*Marine Ecology Progress Series* 96 (1993): 259–67.)

20. *Le Marin*, February 22, 1991.

21. *Conservation of endangered reef building corals in the Arabian Gulf, proposal for immediate and short-term actions by the European Oceanographic Observatory*, March 1991.

22. Corals belonging to the Scleractinia are noted in the Convention on International Trade in Endangered Species (CITES), signed March 3, 1973, in Washington. Highly protected at the time when they were imported to Monaco (the scleractinian corals were listed in Appendix I of the convention), they were delisted in October 1989 (Appendix II).

23. *Science et Vie*, May 1991.

24. J. W. Readman et al., "Oil and combustion-product contamination in the Gulf marine environment following the war," *Nature*, no. 358 (1992): 662–63.

25. See note 17.

26. *Nice-Matin*, February 18, 1991. On the "prud'homie," see note 4.

27. *L'Express*, April 11, 1991.

28. Ibid.

29. "*Caulerpa taxifolia* is more a bio-indicator of the warming of the Mediterranean. It's a question of a 'dormant' species, long established in this habitat, suddenly reactivated by a climatic phenomenon" (*Le Méridional*, January 26, 1992).

30. *L'Express*, April 11, 1991.

31. *Var-Matin*, July 28, 1990 and September 14, 1990.

32. IFREMER is a public entity with an industrial and commercial mission.

33. He had a hundred international publications on this subject to his
credit.

34. *France-Soir,* August 7, 1991.

35. *Le Provençal,* August 5, 1991.

36. *Var-Matin,* August 5, 1991.

37. Ibid.

38. *France-Soir,* August 7, 1991.

39. *Le Monde,* August 14, 1991.

40. *Le Figaro,* August 9, 1991.

41. *Var-Matin,* August 9, 1991.

42. *L'Express,* April 11, 1991; *Les Nouvelles Calédoniennes,* August 10, 1991.

43. *Le Monde,* August 14, 1991.

44. In one of my publications (*Tethys* 4 [1972]: 843–58) is the history of the evolution of populations of *Caulerpa prolifera* on the French continental coast; the hypothesis advanced by Doumenge for its regression is by no means the only possible one.

45. Doumenge: "My hypothesis is that this *Caulerpa* has always existed in the Mediterranean. It had greatly regressed from 1960 to 1979, when the winters were cold, and it has returned with the mild winters" (*Les Nouvelles Calédoniennes,* August 10, 1991).

46. *Les Nouvelles Calédoniennes,* August 10, 1991.

47. *France-Soir,* August 7, 1991.

48. *Le Monde,* August 14, 1991.

49. *L'Express,* April 11, 1991.

50. *Le Figaro,* August 9, 1991.

51. *Les Nouvelles Calédoniennes,* August 10, 1991.

52. Ibid.

53. *Agence France-Presse* dispatch, August 8, 1991.

54. FR3, August 10, 1991.

55. *Les Nouvelles Calédoniennes,* August 10, 1991.

56. Dispatch of August 8, 1991.

57. Letter of Pierre Papon, president and chief executive officer of IFREMER, September 10, 1991: "We are not conducting research specifically on this alga. We have never said that we made use of analytic results as was indicated in the dispatch. We are not responsible for the suggestions attributed to Mr. François Madelain, which were misinterpreted."

58. Roger Cans, *Tous verts* (Everyone is green) (Paris: Calmann Lévy, 1992).

59. This underwater scooter, run by powerful batteries, can operate for an hour, which allows one to traverse nearly 1,500 meters fairly easily.

60. Center for studies of algae and economic development based on them.

61. In 1991, the determination of the area totally covered by *Caulerpa taxifolia* became more and more difficult. Large areas were entirely covered and surrounded by zones in which the colonies were numerous but scattered. We delineated the areas in which the great majority of colonies were concentrated. It is this "affected area" that is used to compare the situation from year to year.

62. A. Meinesz, B. Hesse, and X. Mari, "Situation des zones atteintes par l'algue *Caulerpa taxifolia* sur la Côte d'Azur," Report of the laboratory of the marine coastal environment, University of Nice–Sophia Antipolis, November 15, 1991.

63. Letter of November 18, 1991.

64. Letter of December 2, 1991.

65. Since July 1990, the month when the announcement of the invasion of Cap Martin by *Caulerpa taxifolia* was publicized, we had observed a growing number of private divers. They gathered significant amounts of *Caulerpa taxifolia* that they subsequently transported in garbage cans full of seawater furnished with aerators. This tropical *Caulerpa* generated a fruitful commerce among aquarists. We had indicated the mother lode! Sprigs of *Caulerpa taxifolia* fetched thirty francs. The growing cultivation of the alga could lead to new dissemination "accidents." It was necessary to control this potential source of secondary contamination.

66. "Sauve qui veut" is a pun, based on the common expression, "Sauve qui peut," which is approximated by "Run for your life!" or "Every man for himself!"

67. *Nice-Matin,* December 5, 1991.

68. *Var-Matin* and *Nice-Matin,* December 5, 1991.

69. Letter of October 30, 1991: "It isn't possible for me to authorize the use of this hall for meetings that can produce polemics in the presence of the press and that could embarrass the university in its interactions with its partners."

70. Mrs. Martine Daugreilh, Deputy.

71. Campaign subsidized by the Regional Council of Province–Alpes–Côte d'Azur and the General Council of Alpes-Maritimes.

72. He was subsequently to become the official liaison between the Coordinating Committee on the alga and the directorate of the Oceanographic Museum.

73. Doumenge had just written to the Science Attaché of the French Embassy in Japan to advise him to consult Mr. Merckelbagh to send a Japanese researcher (on a detour!) to an IFREMER branch at Brest (1,500 kilometers from Nice) when the researcher had wanted to spend a sabbatical year in my laboratory to work on *Caulerpa taxifolia.*

74. Order of July 19, 1988; *Journal Officiel* of August 9, 1988.

75. Letter of November 18, 1991.

CHAPTER THREE

1. *Nice-Matin,* December 17, 1991.

2. *Agence France-Presse,* dispatch of December 16, 1991.

3. *Agence France-Presse,* dispatch of December 17, 1991.

4. A student at the School of Mines in Paris submitted a master's thesis on the *Caulerpa taxifolia* affair. He analyzed the scope of the aspects taken up by the press, theme by theme. P. Mazataud, "The construction of scientific truth: The alga *Caulerpa taxifolia,*" Paris School of Mines, July 1993.

5. *Agence France-Presse,* dispatch of December 18, 1991.

6. "Nowadays, François Doumenge, the director of the Museum, does not wish to answer questions about the alga." *Le Figaro,* December 20, 1991.

7. *Agence France-Presse,* dispatch of December 17, 1991.

8. *Le Figaro,* December 20, 1991.

9. *Le Provençal,* January 21, 1992.

10. *Le Monde,* August 14, 1991.

11. *Agence France-Presse,* Paris, dispatch of December 17, 1991.

12. J.-L. Nothias, *Le Figaro,* December 20, 1991.

13. *Apnéa,* March 1991.

14. This adjective was already used to designate dominant species that eliminate their relatives or threaten the existence of other species: killer bees, killer flies, etc.

15. For example, during the summer of 1992, a small insect, *Frankliniella occidentalis,* that locally affected flower production by sucking plant sap, was baptized the "Dracula of flowers" by *Nice-Matin.*

16. "Antibes-Juan-les-Pins. The diving club can breathe easily. *Taxifolia* is not there" (*Nice-Matin*, March 23, 1994); "Cavalaire. Diving: a tourist attraction" (*Le Var-Nice-Matin*, Sainte-Maxime Publ., Saint-Tropez, April 1, 1994); "Cavalaire. A marine observatory" (*Le Var-Nice-Matin*, July 5, 1994).

17. *Le Méridional*, June 23, 1992.

18. Other affairs dealing with global environmental threats—the greenhouse effect, the ozone hole, acid rain—have seen similar developments. See P. Roqueplo, *Climats sous surveillance: Limits et conditions de l'expertise scientifique* (Climate under surveillance: Limits and conditions of scientific expertise) (Paris: Économica, 1993); J.-M. Salles, "The economic stakes of global threats to the environment," *Natures Sciences et Sociétés* 1–2 (1993): 108–17.

19. See chapter 2, note 4.

20. This position is indicated in the minutes of the meeting drafted by Mr. Aquila.

21. "Saint-Cyprien: *Caulerpa taxifolia*, the green cancer 'attached' on the harbor bottom?" *L'Indépendant de Perpignan*, January 16, 1992.

22. "Saint-Cyprien: No fluorescent green alga." *L'Indépendant de Perpignan*, January 25, 1992. "*Taxifolia*: Saint-Cyprien liberated. As a precautionary measure, professional divers will conduct a systematic survey every month in all confined areas at depths where the alga could return." *Le Marin*, February 7, 1992. In fact, two years later, 850 square meters covered by the alga were discovered at a turbid, muddy site; the eradication had been incomplete and the monitoring effort insufficient.

23. "The killer alga has crossed the Rhône." *Nice-Matin*, December 24, 1991.

24. Calcareous agglomerations produced by animals or plants.

25. We had created this certificate to show our support for and recognition of volunteer divers who provided precious assistance by reporting the presence of the alga.

26. "Toxic alga attacks Les Lecques." *Le Provençal*, January 21, 1991.

27. "First green spots at the bottom of the deep blue sea; *Caulerpa taxifolia* has just been spotted at Saint-Cyr. Professor Meinesz sounds the alarm." *Var-Matin*, January 21, 1992.

28. *France Soir*, January 24, 1992.

29. *Le Quotidien de Paris*, January 29, 1992.

30. *Nice-Matin*, January 25, 1992.

31. *La Marseillaise*, January 29, 1992.

32. *Var-Matin*, January 25, 1992.

33. *Nice-Matin*, January 25, 1992.

34. As scientific advisor to the regional federation of associations for environmental protection (URVN: Union régionale pour la protection de la vie, de la nature et de l'environnement = Regional Association for the Protection of Life, Nature, and the Environment).

35. "The Museum counter-attacks." *Le Méridional*, January 26, 1992.

36. Professor Michel Denizot, "In the matter of a *Caulerpa* (green alga) present in the Mediterranean Sea," *Annales de la société d'horticulture et d'histoire naturelle de l'Hérault* 132, no. 3 (1992): 29: "Deliberate dissemination: the discovery of substrates assumed to come from aquaria has even forced the municipality to place the matter in the hands of the courts." This note was distributed by the Museum on May 2, 1992.

37. C.-F. Boudouresque, M. Verlaque, D. Pesando, F. Pietra, F. Cinelli, and myself. I was selected to chair this committee.

38. He allowed himself to be co-opted in this way. This was announced at his meeting and mentioned in a motion widely distributed beginning December 23, 1991.

39. *Agence France-Presse*, Nice, dispatch of December 17, 1991; *L'Union*, December 19, 1991.

40. This communiqué was published in part in *Quotidien de Paris*, December 25, 1991; *Le Midi Libre*, December 26, 1991; *Le Méridional*, December 26, 1991; and *Var-Matin*, December 28, 1991.

41. *Le Pêcheur Varois*, December 1991.

42. Letter entitled, "Request for the formation of a coordinating committee to battle the alga *Caulerpa taxifolia*." A copy of this letter was addressed to all concerned ministries and to regional political authorities.

43. Research totaling 300,000 francs.

44. This latter individual never attended the meetings because the administration of Maritime Affairs, which convened the committee meetings, was unable to guarantee his travel expenses!

45. In 1952, Alain Bombard became the first person to cross the Atlantic in a small inflatable boat, without a motor and without provisions. His voyage lasted 113 days, and he survived by eating fish, seabirds, and plankton (scooped up with a small net) and by drinking less than 800 grams of seawater a day. A physician, he knew the amount of seawater a person could tolerate. He proved that a person can survive on the ocean on very little; although he had emer-

326 gency rations with him, he never touched them. Subsequently a widely cited spokesperson in France for all matters relating to the sea, he served as Minister of the Environment in France and as deputy in the European Union.

46. This sentiment was shared by the reporters present: "Killer alga: War!" *Var-Matin*, January 24, 1992; "Declaration of war on the killer alga!" *Le Méridional*, January 24, 1992; "Killer alga: Mobilization. Toxic alga: It's war." *Le Provençal*, January 24, 1992.

47. *Le Progrès de Lyon*, January 25, 1992.

48. Comprising thirty-two members, of whom fourteen represented major national government agencies (prefects, departmental or regional directors of Maritime Affairs or of equipment), four represented local communities, five represented fishermen, three were IFREMER experts, one represented divers and yachtsmen, and four were scientists: Professors Boudouresque, Vicente, and Meinesz, and Mrs. Pesando of INSERM.

49. Tintin is a famous comic strip character in Belgium and France, created by Hergé. His adventures have been translated into more than forty other languages.

50. The photograph appeared in *Sciences et Avenir* in March 1992, with this interpretation of the affair attributed to Jaubert: "The current explosion reflects the reawakening of a dormant species, stimulated by the mild winters and growing water pollution."

51. Lucien Chabason, advisor to the cabinet of Brice Lalonde and head of the electoral list for the party Génération Écologie in Var, had just announced the week before at his office in Toulon that his ministry "had decided to free up the sum of 500,000 francs to initiate the first research on *Caulerpa taxifolia*" (*Var-Matin*, January 21, 1992). These funds, promised by the Ministry of the Environment in July 1992, did not reach the laboratories until January 1995, 21 months after the arrival of Michel Barnier at the Ministry of the Environment.

52. *Nice-Matin*, February 1, 1992.

53. The deputy from Var, Mrs. Yann Piat (who fell to killers' bullets in 1994) posed a written question to the Ministry of the Environment (question no. 53378 of January 27, 1992); José Balarello, senator from Alpes-Maritimes, requested an administrative inquest to determine responsibilities (*Nice-Matin*, February 20, 1992); Paul Lombard, deputy from Var, posed a written question to the Ministry of the Environment (question no. 52989 of January 20, 1992); René-Georges Laurin, senator and mayor of Saint-Raphaël, sent a formal notice

to the Ministry of the Environment to undertake the eradication of the colony   **327**
discovered in his community (*Nice-Matin*, January 22, 1992).

54. Région Verte, association of elected officials from various communities, announced it was lodging a judicial complaint (against an unknown person) at the court at Grasse for introducing a foreign species (referring to a coastal law) (*Nice-Matin*, January 31, 1992); the prud'homie of Bastia also announced a legal maneuver in the local press: "Complaint of fishermen against the Oceanographic Museum" (*La Corse*, February 19, 1992).

55. To my knowledge, there has been no sequel to the these complaints.

56. Jaubert had been appointed director of the Center of Marine Biology on the faculty of sciences at the University of Nice in 1991. This center comprised two small marine zoology laboratories located on the same floor as mine. Despite this reorganization, he conducted all his research at the Oceanographic Museum of Monaco, where he had been named director of a laboratory of Monacan research (the European Oceanographic Observatory). At the end of 1992, he asked the national education authorities to second him to that position.

57. For our operations and capital investments we were able to count on 30,000 francs per year for the four teaching-researchers in my mini-laboratory. In fact, it was not this modest sum that was important: the recognition of the status of our team enabled us to host advanced students in good circumstances, and this was necessary to support the various research projects undertaken by our team.

58. L. Mayet, "Green alert in the Mediterranean," *Sciences et Avenir,* March 1992.

59. J. Artaut, "Say it with algae," *Aquarium Magazine* 26 (1987): 39–42.

60. *Libération*, February 19, 1992.

61. "Monaco: the Museum counter-attacks." *Le Méridional,* January 26, 1992.

62. April 13, 1992, at the University of Nice-Sophia Antipolis.

63. "Attempt to implant *Caulerpa prolifera*," *Compte rendu des activités de l'Association monégasque pour la protection de la Nature*, 1992–93, 19.

64. "Everyone is responsible for the damage he causes not only by his actions but also by his negligence or carelessness," article 1,383 of the Civil Code, cited in Professor Jean Dorst's book, *Avant que nature se meure* (Paris: Delachaux and Niestlé, 1978), 285.

65. Four years earlier, Cousteau's foundation had asked me for an opinion piece on the health of the Mediterranean. The article, which appeared in the

328 magazine *Calypso Log* (February 1985), was reprinted in the book by J.–Y. Costeau and Y. Paccalet, *La mer blessée: La Méditerranée* (Paris: Flammarion, 1987). Cousteau wrote an introduction to my article and also asked me to collaborate in illustrating the work.

66. *Le Point,* February 15, 1992.

67. *Le Figaro,* February 25, 1992.

68. *Libération,* February 19, 1992.

69. *Nice-Matin,* February 20, 1992.

70. *Nice-Matin,* February 24, 1992.

71. *Le Parisien,* March 2, 1992.

72. In *Nice-Matin* on February 24, 1992, the "next update by Commander Cousteau" was reported with this commentary: "Present Saturday evening in Monaco in the course of a strictly private visit, Commander Cousteau refused to discuss, for the moment, the problem and the different positions on the origin of the spread of the 'killer alga' in the Mediterranean. According to reliable sources, the foundation that carries his name is going to produce an ecological balance-sheet, and, when it is issued, the celebrated oceanographer is going to issue a summary statement that is eagerly awaited by all parties and that will not be delayed."

73. Editorial of Jean Attard, director of *Apnéa,* monthly publication for underwater fishermen, March 1992.

74. D. Ody, "The irresistible invasion of *Caulerpa taxifolia,*" *Calypso Log,* April 1992.

75. B. Violet, *Cousteau* (Paris: Fayard, 1992), 267–69.

76. "This story has been blown up out of all proportion. It is part of a larger problem, the movement of viruses, microbes, insects, etc. outside their habitual ranges because of the speed of modern means of transportation. What is currently happening on the banks of the Mediterranean appears to me neither tragic nor definitive. There was a precedent in the 1950s with another algal introduction, this time of a brown alga. And then everything returned to normal. Life changes . . . Nature changes . . . It is very difficult for an honest ecologist to distinguish the things that are truly serious from those that are ephemeral." (*Nice-Matin,* October 20, 1992)

77. D. Ody, "The inexorable progress of *Caulerpa,*" *Calypso Log,* June 1994.

78. *Le Point,* February 15, 1992.

79. We have never been able to verify this scoop; at the time of publication of this book, no publication of our local phycological colleagues records the presence of *Caulerpa taxifolia* in Tunisia.

80. *Le Méridional,* January 26, 1992; *Le Point,* February 15, 1992.

81. *Apnéa,* March 1992.

82. *Apnéa,* March 1992; *Le Figaro,* February 25, 1992.

83. *Le Méridional,* January 26, 1992.

84. Ibid.

85. *Thalassa,* France 3, March 1992.

86. Jean Duclerc, "Invasion of *Caulerpa taxifolia,*" *Compte rendu de plongée au Cap Martin (06) le 18/02/1992,* IFREMER, Sète station, laboratory of fishing resources; Duclerc, "Attempts at manual eradication of *Caulerpa taxifolia,*" IFREMER, Sète station, laboratory of fishing resources (February 14, 1992).

87. At the meeting organized by the Toulon fishermen on December 23, 1991, this expert, based in Sète, had already shown his colors by supporting the "doumengian" hypothesis that the alga came directly from the Red Sea.

88. In his monthly journal, he went so far as to write: "The Coordinating Committee should be the only source of controlled and authenticated information (which is not the case today) on the nature of this alga and its impacts on the environment" (*Le Pêcheur varois,* February 1992.)

89. Mrs. Heymonet had been designated vice-chair of the Coordinating Committee. She represented the Regional Council of Provence–Alpes–Côte d'Azur, of which she was a member. Her term ended in March 1992. Candidate for reelection to that position, she won and was named by her peers chair of the commission on the sea of the Regional Council in June 1992.

90. *Nice-Matin,* February 25, 1992.

91. Three-page note entitled, "Taking stock of the alga *Caulerpa taxifolia,*" edited on IFREMER letterhead paper and signed by Yves Henocque on February 20, 1992.

92. The "poison pen" took the trouble to indicate the addressees of his or her venomous words: "This letter was sent to all entities concerned with the *Caulerpa* problem, as well as to the media."

93. One of the poison pen's letters was sent anonymously to me stamped "received" by three agencies in the same community.

CHAPTER FOUR

1. "*Caulerpa taxifolia:* no panic," *Nice-Matin*, Monaco-Menton edition, August 23, 1992.

2. See chapter 2, note 4.

3. "*Caulerpa* reaches the Bay of Sablettes. The alga covers all the sites where fish reproduce," *Nice-Matin*, Monaco-Menton edition, October 27, 1994.

4. *Nice-Matin*, February 2, 1991.

5. Minutes (approved April 10, 1995) of the meeting on *Caulerpa taxifolia* of the committee on the sea of the economic and social council of the Provence–Alpes–Côte d'Azur region, organized by IFREMER on December 21, 1994.

6. "Killer alga or cancer of the sea. Worry among fishermen of the far south," *La Corse*, February 6, 1992.

7. "Attack on *Caulerpa taxifolia.* Complaint lodged by fishermen against the Oceanographic Museum." Under this headline, the article indicated that the fishermen of Bastia asked their lawyer "to summon the Oceanographic Museum of Monaco before the court. This step will be accompanied by a suit for damages." *La Corse*, February 19, 1992.

8. "If it arrives here and if no one does anything, the truckers' blockade will be nothing compared to what we will do. . . . The explosion of their anger will be terrifying, predict the mayors of the affected communities" (*Méditerranée magazine*, September 1992).

9. "The minister congratulates the divers," *Subaqua*, July–August 1992.

10. P. Mouton, "200 beautiful Mediterranean dives (the French coast)," *Océans* (special publication), editions of 1973, 1981, 1984.

11. J. Artaut, "Say it with algae," *Aquarium Magazine* 26 (1987): 39–42.

12. Until 1984, the president of UCA was Michel Hignette, head of the aquaria of the Museum of Monaco.

13. Pierre Escoubet: "We must not lose any more time. Either we destroy it now or the situation will become hopeless if the invasion worsens." *Agence France-Presse* Nice, July 12, 1992; *Le Quotidien de Paris*, July 13, 1992.

14. *Aquarama*, March 1992.

15. M. Tassigny, "Concerning *Caulerpa taxifolia*," *Subaqua*, May–June 1992.

16. J.-P. Alayse, "*Caulerpa taxifolia*, a killer alga?" *Océano news* (journal for members of Océanopolis), no. 1 (April 1992).

17. The order was to be signed by the Minister of the Environment at the

time, Ségolène Royal: ministerial order of March 4, 1993; *Journal Officiel* of March 25, 1993.

18. "*Caulerpa taxifolia* banned! Aquarists targeted," *Aquarium Magazine,* November 1993.

19. "Petition against the order banning *Caulerpa taxifolia*" (proposed by the editorial board of the magazine), *Aquarium Magazine,* December 1993.

20. See chapter 3, note 45.

21. A. Apoteker, "Mediterranean: alert for the killer alga," *Greenpeace,* June–July–August 1992.

22. Press release by Greenpeace, Palma, Mallorca, October 28, 1992.

23. "*Caulerpa* is continually extending its empire: can we still fight it?" *Naturellement* (journal of the national movement to fight for the environment), Spring 1994; "The expansion of *Caulerpa taxifolia* continues," *Le Courrier de la Nature,* May–June 1994.

24. A. Meinesz, "*Caulerpa taxifolia,* the evil flower," *Planète Mer,* October–November-December 1994; A. Meinesz, "*Caulerpa taxifolia,* an evil flower," *Espaces pour Demain,* 1st trimester 1995.

25. "Killer alga in the Mediterranean: the government sticks its head in the sand," *La Dépêche de l'Environnement* (bulletin of the ecological news agency), June 3, 1992.

26. "The Rotary Clubs listen to the needs of the community," *Nice-Matin,* November 13, 1994; "Rotary International. *Taxifolia*—known as the killer alga—theme of Professor Doumenge's lecture," *Nice-Matin,* Antibes-Vallauris-Cannes edition, December 3, 1995.

27. A. Meinesz, "The alga *Caulerpa taxifolia* continues its invasion of the Mediterranean," *Le Rotarien,* July 1996.

28. "I have the impression that the national government is playing with the clock and trying to avoid spending money." Remark at a round table organized at the Marseilles Aquarium. *Aquaforum,* June 5, 1992; "Killer alga: Professor Meinesz's warning call," *Le Provençal,* June 6, 1992.

29. Alerted in November 1991.

30. Recipients of my plan in January 1992.

31. J. Duclerc, "Attempts at manual eradication of *Caulerpa taxifolia,*" IFREMER, Sète station, February 14, 1992; "In pursuit of the killer alga," *Nice-Matin,* February 12, 1992; "Toxic alga: The French Navy enters the fray," *Le Soir,* February 12, 1992.

32. *Le Var* and *Nice-Matin,* February 13, 1992.

332   33. "Attack on *Caulerpa taxifolia*," *Nice-Matin*, Monaco-Menton edition, March 9, 1992.

34. "In any event, a total eradication is impossible and the attempt is certainly not justified. It is necessary first of all to work on the key points: the *Posidonia* meadows" (Y. Henocque, of IFREMER, cited in *Libération*, February 19, 1992); "Dr. Yves Henocque, head of the project at IFREMER, is under no illusion: 'I no longer believe eradication is possible. When a species grows so rapidly, there is hardly any hope of eliminating it totally. At most we can drive it out of particularly sensitive zones like the meadows or harbor entrances'" (*Le Point*, February 15, 1992).

35. *Apnéa*, March 1992.

36. The operation was carried out while I was in Japan, invited by research institutions to lecture on the impact of coastal zone development on marine ecosystems.

37. *Nice-Matin*, January 22, 1992.

38. "Railing against *taxifolia*," *Le Var* and *Nice-Matin*, March 4, 1992; "Saint-Raphaël: railing against *taxifolia*," *Nice-Matin*, March 4, 1992; "*Taxifolia*: the battle is joined; Killer alga: the counterattack is planned," *Nice-Matin*, March 6, 1992).

39. *Nice-Matin*, March 6, 1992.

40. "The eradication market is valued at several tens of millions of francs (and is growing every year), which gives rise to a severe competition between companies. Many have already joined the battle" (*Le Méridional*, June 11, 1992).

41. Official decision no. 23/92 of the Maritime Prefecture of the Mediterranean.

42. "Killer alga: eradication tests planned for May at Cap Martin," *Nice-Matin*, April 29, 1992; "Weeding the sea," *La Croix*, June 12, 1992; "The toxic alga in the Mediterranean: should we wait or eradicate?" *Le Quotidien du Médecin*, June 23, 1992.

43. "The alga *Caulerpa taxifolia*. Comparative test of different eradication techniques," *High-Tech Environnement*, July 1992.

44. "The representative of the Interministerial Committee for the Sea called to the attention of the main forces behind the great crusade against the 'killer alga' that objective data on the behavior and danger of *Caulerpa taxifolia* in the western Mediterranean are still wanting. In short, researchers have not

always provided scientific proof of their discoveries" ("Green alga: science in   **333**
troubled waters," *Le Méridional,* July 25, 1992).

45. "Young shoots torn out," *Le Var-Nice-Matin,* August 22, 1992.

46. "*Caulerpa* tested by chlorine and ultrasound," *Nice-Matin,* July 21, 1994.

47. "A predator serves science. Seahare against killer alga," *Nice-Matin,* April 12, 1992.

48. Above a photograph of the mayor of Marseilles contemplating a jar containing *Aplysia,* the caption reads: "*Aplysia punctata* obviously fascinated Robert Vigouroux." *Le Méridional,* April 15, 1992.

49. *Nice-Matin,* December 5, 1991.

50. On two occasions in January 1992, Mr. Henocque of IFREMER mentioned the possibility of quicklime.

51. Following the operation conducted by IFREMER at Toulon in which the alga was manually ripped out, the press reported: "As for the chemical procedure, it seems to have been rejected by IFREMER." *Nice-Matin,* February 12, 1992. "Obviously concerned about pollution, IFREMER has rejected the use of quicklime envisaged at one time by certain experts." *Le Soir, Le Provençal* edition, February 12, 1992.

52. "Professor Boudouresque believes that we should not preclude the use of any technique; even chemical ones and those that appear at first blush to be the most eccentric should not be eliminated from the experimentation program" (*Compte rendu de la première rèunion du comité de coordination sur l'algue Caulerpa taxifolia,* Nice, February 24, 1992); "Lime has already been successfully tested in the United States and does not pollute at all, because it decomposes into calcium carbonate, emphasized Professor Boudouresque" (*Le Quotidien de Paris,* March 9, 1992).

53. See chapter 3, pp. 82–88.

54. Bernard Jaffrennou and Lucien Oddone.

55. "Toulon: victory over the killer alga," *Var-Matin,* August 5, 1993; "Toulon: the killer alga doesn't like copper," *Nice-Matin,* August 6, 1993.

56. A reception was organized on October 8, 1993, at the regional offices to present the procedure.

57. CEV = cuvée électrolytique virtuelle = virtual electrolytic vat.

58. "*Caulerpa* tested by chlorine and ultrasound," *Nice-Matin,* July 21, 1994.

59. Conversion, Limited, which has a factory at La Ciotat.

60. Laboratory of membrane materials and methods. Joint research unit of CNRS, the University of Montpellier I, and the National Advanced School of Chemistry.

61. Team directed by Claude Gavach, research director at CNRS.

62. "Success for four Montpellier researchers. How they have conquered the 'killer alga,'" *Le Midi Libre*, September 23, 1994; "The killer alga vanquished? An effective weapon against *Caulerpa*," *L'indépendant de Perpignan*, September 22, 1994; "Anti-*taxifolia* membrane," *Le Nouvel Observateur*, October 8, 1994; "A weapon tested against the foreign alga," *Le Figaro*, September 24, 1994; "The serial of *Caulerpa taxifolia*," *Science et Vie*, December 1994.

63. "*Caulerpa taxifolia*, play it again," *Science et Vie*, February 1995; "A robot assaults the toxic alga," *Var-Matin*, October 27, 1994.

64. "Sea salt against the 'killer alga,'" *Agence France-Presse*, January 15, 1996; *La Dépêche*, January 16, 1996; "Sea salt kills *taxifolia*," *Nice-Matin*, January 16, 1996; "Toxic alga: the salt of hope," *Var-Matin*, January 22, 1996; "Salt *Caulerpa*," *Le Figaro*, January 17, 1996.

65. Drawing presented on FR3 Côte d'Azur on January 18, 1996.

66. Press release of July 29, 1992.

67. In the hundreds of hectares then infested, the maximum measured biomass surpassed ten kilograms/square meter.

68. "Yacht anchorage will be changed in the Mediterranean to contain the spread of a tropical alga," *Le Monde*, October 22, 1992.

69. "Killer alga in the Mediterranean: the government sticks its head in the sand," *La Dépêche de l'Environnement*, June 3, 1992.

70. P. Papon, "The politics of IFREMER. IFREMER and its assessment mission," *IFREMER information*, March 1993.

71. *IFREMER information*, September 1992.

72. *Subaqua*, July–August 1992.

73. M. G. Monnier-Besombes, *Threat to the Mediterranean coast: Caulerpa taxifolia*, 1993 (brochure produced by the environmental working group of the green party in the European Parliament).

74. The promotion was arranged by the Ministry of Tourism. *Journal officiel* of April 11, 1993.

75. M. Barnier, *Atlas of Major Threats* (Paris: Plon, 1993), 74.

76. Deputy Suzanne Sauvaigo, then mayor of Cagnes-sur-Mer, twenty   **335**
kilometers from Nice.

77. D. Ody, "The inexorable progress of *Caulerpa*," *Calypso Log*, June
1994.

78. This was also the conclusion of an analysis of the affair conducted by
engineering students of the National School for Engineering of Rural Waters
and Forests (ENGREF = École nationale du génie rural des eaux et forêts):
D. Charignon, F. Dupont, T. Tiengou, and C. Varret, "*Caulerpa taxifolia:* study
of the sites and proposals for a management program," Report TGE, ENGREF,
Paris, 1996.

79. *La Revue du Golfe de Saint-Tropez*, August–September 1992.

80. In France, the region Provence–Alpes–Côte d'Azur was the principal
funding source for research on *Caulerpa taxifolia:* this region provided 750,000
francs from 1990 to 1996.

81. Minutes of the regional marine committee, June 16, 1993.

82. Patrick Allemand, note to elected officials of coastal communities,
September 28, 1993.

83. "*Caulerpa* worries the Regional Council. To battle against the 'killer
alga,' the regional marine committee wants a charter of the nations of the Mediterranean basin," *Nice-Matin*, September 20, 1995.

84. However, some people have manifested their concern, such as Bernard
Deflesselles, who published a realistic editorial in the newsletter of his association: "*Caulerpa:* infinitesimal cause, enormous consequences," *Lettre de l'association environnement et société*, January 1993.

85. The Observatory of Mediterranean Fauna and Flora.

86. Meeting held under the aegis of the mayor of Menton, Jean-Claude
Guibal (cf. "Killer alga: Politicians mobilize," *Nice-Matin*, February 16, 1992).

87. During the last cruise, a team of IFREMER filmmakers shot a film.
The film was subsequently commercialized by IFREMER with the title, "On
the trail of the Caulerpe" (IFREMER, July 1995).

88. *Recherches marines [Marine Research]*, IFREMER, July 1995.

89. "When there are threats of serious or irreversible damage, lack of full
scientific certainty shall not be used as a reason for postponing cost-effective
measures to prevent environmental degradation." "*Caulerpa taxifolia:* Biodiversity in danger?" *Recherches marines*, IFREMER, July 1995.

90. *L'Action universitaire*, June 1994.

336   91. He was named by the Minister of Education, Higher Instruction, Research, and Professional Integration to the National Committee on Scientific Research, Section 30 (decree of October 3, 1995; *Journal officiel* of October 21, 1995).

92. *Nice-Matin*, Menton-Monaco edition, February 16, 1992.

93. "Killer alga arrives at Imperia," *La Stampa*, May 31, 1992.

94. "Green alert," *Mondo sommerso*, June 1993; "The killer from the tropics," *Airone*, August 1993; "Searching for *Caulerpa taxifolia*," *Aqva scienza magazine*, September 1993; "Alert in Sicily. The killer alga has crossed the straits," *Vivere La Sicilia*, October 1993; "Story of an alga," *Il Subacqueo*, March 1995.

95. Fulvio Grimaldi, scientific commentator of *Rai 3*, producer of a program on *Caulerpa taxifolia* broadcast on November 19, 1994.

96. ENEA (National Committee on Research and Development of Nuclear and Alternative Energies) and the Inter-university Center for Marine Biology of Leghorn (CIBM).

97. "Killer alga: Fishermen launch an SOS," *La Stampa*, August 24, 1993.

98. "*Caulerpa taxifolia* in the straits? Yes, the presence of the 'killer alga' is confirmed by biologists," *La Sicilia*, July 21, 1993.

99. Famous whirlpool and reef of the Straits of Messina immortalized in mythology as monsters lying in ambush for sailors in distress. The pilot, if he escapes one danger, is sure to be caught by the other. This is the origin of the proverb, "to go from Charybdis to Scylla," equivalent of "to go from bad to worse."

100. "Hunting the killer alga; An expert: alarm is unwarranted" (*Giornale di La Sicilia*, August 6, 1993); "But the alga is not a killer" (*La Sicilia*, August 6, 1993).

101. Nico Orengo, *La Guerra del Basilico* (The War of the Basil) (Turin: Guilio Einaudi, 1994).

102. Decree of the President of the Region of Catalonia, no. 257, October 26, 1992.

103. Letter of His Royal Highness Prince Rainier of Monaco, December 14, 1991.

CHAPTER FIVE

1. A large part of the scientific work is found in the proceedings of two symposia, one held at Nice (January 1994), the other at Barcelona (December 1994): C.-F. Boudouresque, A. Meinesz, and V. Gravez, eds., *First International Workshop on Caulerpa taxifolia* (Marseilles: GIS-Posidonie, 1994); M. A. Ribera, E. Ballesteros, C.-F. Boudouresque, eds., *Second International Workshop on Caulerpa taxifolia* (Barcelona: Publicacions de la universitat de Barcelona, 1996). An inventory of 161 scientific documents on the alga is also found in C.-F. Boudouresque, A. Meinesz, and V. Gravez, *Travaux scientifiques consacrés à l'algue introduite en Méditerranée, Caulerpa taxifolia* (Marseilles: GIS-Posidonie, 1996).

2. The molecule has also drawn the attention of physicians; a team from Nice has shown that it blocks division of cancer cells without killing them.

3. "Barnier Law": no. 95–101, February 2, 1995, *Journal officiel*, February 3, 1995.

4. *United Nations Environmental Program. Action Plan for the Mediterranean* (Athens: UNEP, March 3, 1995).

5. Recommendation of permanent committee no. 45, March 21, 1995.

6. *United Nations Environmental Program. Action Plan for the Mediterranean* (Athens: UNEP, March 3, 1995).

7. The CIESM, based in Monaco, was presided over by Prince Rainier, and, at that time, Professor Doumenge was secretary-general.

8. "'Killer alga' divides scientists," *Le Provençal*, July 30, 1992.

9. "Duel for the killer alga," *Le Quotidien de Paris*, August 1, 1992.

10. "'At the Observatory,' states Professor Jaubert, 'we are studying the factors controlling the growth of *Caulerpa taxifolia*. The first data clearly indicate that it is not an invasion: the several dozen hectares of underwater bottom in question have scattered small colonies between Italy and Spain'" (*Nice-Matin*, Monaco-Menton edition, July 12, 1992).

11. Direct summons before the correctional court of Paris, August 14, 1991.

12. *Nice-Matin*, Monaco-Menton edition, July 12, 1992.

13. *Le Méridional*, July 30, 1992.

14. Ibid.

15. Jaubert's former laboratory and mine were located on the 4th floor of the natural science building of the faculty of sciences at the University of Nice.

16. Document concerning the assessment of my laboratory, which led one

338   to believe incorrectly that it had not been granted the status of "laboratory permitted to host graduate students," unlike that of Jaubert.

17. F. Jubelin, "*Caulerpa* madness," *Thalassa,* December 1992–January 1993.

18. "Monaco: the Oceanographic Institute slandered. *Envoyé spécial* condemned for an alarmist report on the alga *Caulerpa taxifolia*," *Nice-Matin,* February 12, 1993.

19. Appeals court of Paris, June 23, 1993.

20. "*Caulerpa* reactions. Reply of Alexandre Meinesz," *Thalassa,* April 1993.

21. "Professor Doumenge: 'I declare to you that not only is *Caulerpa taxifolia* clearly declining in Monacan waters, but that this species is also contributing to the return of living organisms in sites where the biotic community has been destroyed by pollution'" (*Le Figaro,* September 24, 1994).

22. R. Tardy, *Var-Matin,* February 16, 1995.

23. "At a time when France is pursuing a moratorium on nuclear tests, the Atomic Energy Commission is studying, with Jean Jaubert and François Doumenge, director of the Oceanographic Museum of Monaco, the possibility of rehabilitating the reefs of Mururoa Atoll, damaged by the blasts. The method would entail transplanting corals produced by methods developed in the Principality" (M. Mennessier, "Corals finally tamed," *Science et Vie,* August 1993.) "The French government is looking to both Jean Jaubert and François Doumenge to help in the reconstruction of the Mururoa Atoll in the south Pacific, destroyed by continuous French nuclear testing" (*Blue Coast,* November 1993); "In the Pacific, he [Professor Jaubert] is collaborating with the Atomic Energy Commission to reestablish the coral reefs degraded by the disturbance caused by nuclear tests" (*Carnet de Route des hautes technologies,* July-August 1994).

24. "Effect of light, temperature and nutrients on the growth rate and productivity of *Caulerpa taxifolia:* program implemented in Monaco by the European Oceanographic Center."

25. Letter from His Royal Highness Prince Rainier, January 13, 1994.

26. P. Gayol, C. Falconetti, J. Chisholm, and J. Jaubert, "Effect of light, temperature and nutrition on *Caulerpa taxifolia* photosynthesis and respiration," C.-F. Boudouresque et al., eds., *First International Workshop on Caulerpa taxifolia* (Marseilles: GIS–Posidonie, 1994), 291–93.

27. A. Guerriero, F. Marchetti, M. D'Ambrosio, S. Senesi, F. Dini, and

F. Piera, "New ecotoxicologically and biogenetically relevant terpenes of the **339** tropical green seaweed *Caulerpa taxifolia* which is invading the Mediterranean," *Helvetica Chimica Acta* 76 (1993): 855–64.

28. A. Guerriero, A. Meinesz, M. D'Ambrosio, and F. Pietra, "Isolation of toxic sesqui- and monoterpenes from the tropical green seaweed *Caulerpa taxifolia* which has invaded the region of Cap Martin and Monaco," *Helvetica Chimica Acta* 75 (1992): 689–95.

29. M. de Pracontal, *L'Imposture scientifique en dix leçons* (Scientific Fraud in Ten Lessons) (Paris: La Découverte, 1986).

30. G. Giaccone and V. Di Martino, "Les Caulerpes en Méditerranée: un retour du vieux bassin de la Téthys vers le domaine de l'Indo-Pacifique."

31. Responding to the wish of the roundtable organizers to publish the oral presentations, the text appeared the following year with the required changes (G. Giaccone and V. Di Martino, *Biol. Mar.* 2, no. 2 [1995]: 607–12).

32. R. Cans, *Le Monde*, December 21, 1994.

33. *Var-Matin*, July 5, 1994.

34. *Midi Libre*, October 9, 1994.

35. The references for these assertions are given in my article, "Introduction and invasion of the tropical alga *Caulerpa taxifolia* in the northwestern Mediterranean," *Oceanologica Acta* 14, no. 4 (1991): 415–26.

36. A. Meinesz et al., "Preliminary taxonomic notes on *Caulerpa taxifolia* and *Caulerpa mexicana*," C.-F. Boudouresque et al., eds., *First International Workshop on Caulerpa taxifolia*, 105–14.

37. J. Jaubert, "Point of view on the spread of *Caulerpa taxifolia*," Oceanographic Observatory of the Scientific Center of Monaco, June 28, 1994; J. Jaubert, "Commentary on the spread of *Caulerpa taxifolia*," Oceanographic Observatory of the Scientific Center of Monaco, July 1994; "A plea for *Caulerpa*. Jean Jaubert, Professor of Marine Biology at the University of Nice-Sophia Antipolis and specialist in tropical habitats is a defender of *Caulerpa*," *Océans*, October 1994.

38. The arrival of *Caulerpa taxifolia* at Messina had just been described in a scientific paper: C. Fradà-Orestano, *Giorn. Bot. Ital.* 128, nos. 3–4 (1994): 813–15.

39. "The killer alga of the Mediterranean has not hit Sicily, affirms Professor Giaccone," *La Sicilia*, March 14, 1995; "Not only a killer," *Aqua*, August 1995.

40. The term "metamorphosis," never used until then in botany, generally

340   describes substantial changes in form such that the species is no longer recognizable (for example, the transformation from a tadpole into a frog, or from a caterpillar to a chrysalis and then to a butterfly).

41. A. Meinesz and C.-F. Boudouresque, "On the origin of *Caulerpa taxifolia*," *Comptes rendus de l'Academie des Sciences (III, Sciences de la Vie)* 319 (1996): 603–13.

42. In the rebuttal published in the *Comptes rendus de l'Académie des Sciences* (see note 41), the conclusion suggests that "Chisholm, Jaubert, and Giaccone confounded matters by associating recognized and established facts (warming of the Mediterranean, movement of species from the Red Sea to the eastern Mediterranean) with the introduction and spread of one species, the Mediterranean strain of *Caulerpa taxifolia*, for which the biology and chronology of spread are not at all in accordance with these facts" (611).

43. J. Chisholm, J. Jaubert, and G. Giaccone, "*Caulerpa taxifolia* in the western Mediterranean: accidental introduction or migration from the Red Sea," *Comptes Rendus de l'Académie des Sciences (III, Sciences de la Vie)* 318 (1995): 1219–26.

44. Press release, January 3, 1996.

45. Press release of January 15, 1998, signed "Jaubert, Professeur de Biologie Marine, Détaché de l'Université de Nice, Directeur de l'OOE."

46. *Le Figaro*, January 17, 1996.

47. *Le Progrès de Lyon*, February 25, 1996.

48. Correspondence with the editor of *Comptes Rendus de l'Académie des Sciences*.

49. "Metamorphosis paper greeted with derision," *New Scientist*, February 15, 1996.

50. *Le Figaro*, January 17, 1996.

51. *Le Figaro*, July 25, 1996.

52. *Sciences et Avenir*, March 1996.

53. *New Scientist*, February 15, 1996.

54. *Le Figaro*, January 17, 1996.

55. Ibid.

56. In the acknowledgments section of the article on "metamorphosis," the Principality of Monaco and Showboats International are cited for having financed the study. In 1993, Showboats International had sponsored Jaubert's research for a total of $183,000 (*Blue Coast*, November 1993).

57. Dispatch of *Agence France-Press* (Nice branch), February 20, 1996.

58. *Sciences et Avenir*, March 1996.   341

59. *Var-Matin*, February 21, 1996.

60. See note 41.

61. According to Jean Rosa (hematologist and biochemist, secretary-general of *Comptes rendus*), "*Comptes rendus* had not taken sides in what was nothing but a scientific controversy" (*Eurêka*, November 1996); "The nature of the quarrel is a scientific rivalry over important financial stakes" (*Science et Vie*, September 1996). According to François Gros (biochemist and editor-in-chief of *Comptes rendus*), "scientific controversies advance knowledge" (*VSD*, August 8, 1996). According to academician Michel Ladzunski (biochemist of the University of Nice–Sophia Antipolis), "I do not understand why the Academy of Sciences is attacked to this degree. In December 1995, it published the article of Jean Jaubert because it was of scientific interest. From the standpoint of the Academy of Sciences, the two approaches are sound" (*Sophiapolis Riviéra*, October 11, 1996).

62. According to the director of the Max Planck Institute for research in criminal law, it is preferable to sanction abuses internally because all external attacks lead to a retreat into defensiveness ("Fraud: when researchers lose their honor," *Courrier International*, April 4, 1996.)

63. T. D. Center, J. H. Frank, and F. A. Dray, Jr., "Biological control," in D. Simberloff, D.C. Schmitz, and T. C. Brown, eds., *Strangers in Paradise. The Impact and Management of Nonindigenous Species in Florida* (Washington, D.C.: Island Press, 1997), 245–63.

64. L. Civeyrel and D. Simberloff, "A tale of two snails: Is the cure worse than the disease?" *Biodiversity and Conservation* 5 (1996): 1231–52.

65. V. Tardieu, "Méditerranée: la lèpre verte" (The Mediterranean: the green plague), *Science et Vie Junior*, April 1995, 78–83.

CHAPTER SIX

1. These two writings reiterate the hypotheses developed by Jaubert in his publication in *Comptes Rendus* and his note drafted for the Academy of Sciences colloquium.

2. *Science et Vie*, December 1997.

3. Dispatch of November 26, 1997.

4. France 3, November 27, 1997.

5. *Le Figaro*, October 23, 1998.

6. Letter of January 21, 1994; extract published in *Santé Magazine*, July 1997.

7. *Federal Register*, vol. 63, no. 233 (December 4, 1998), pp. 67012–13, and vol. 64, no. 50 (March 16, 1999), pp. 12881–89.

8. O. Jousson et al., "Molecular evidence for the aquariologic origin of the green alga *Caulerpa taxifolia* invading the Mediterranean Sea," *Marine Ecology Progress Series* 172 (1998): 275–80.

9. J. Olsen et al., "Mediterranean *Caulerpa taxifolia* and *Caulerpa mexicana* (Chlorophyta) are not conspecific," *Journal of Phycology* 34 (1998): 850–56.

10. A. Meinesz et al., "Preliminary taxonomic notes on *Caulerpa taxifolia* and *Caulerpa mexicana*," in C.-F. Boudouresque et al., eds., *First International Workshop on Caulerpa taxifolia* (Marseilles: GIS-Posidonie, 1994), 104–14.

11. D. C. Schmitz and D. Simberloff, "Biological invasions: a growing threat," *Issues in Science and Technology* 13 (1997): 33–40.

12. See his book, *The Diversity of Life* (Cambridge, Mass.: Harvard University Press, 1992).

13. E. O. Wilson, "Foreword," in D. Simberloff, D.C. Schmitz, and T. C. Brown, eds., *Strangers in Paradise*, ix–x, at x.

14. A. Meinesz et al., eds., *Suivi de l'invasion de l'algue tropicale* Caulerpa taxifolia *en Méditerranée: situation au 31 décembre 1997*. Publ. Lab. Environnement Marin littoral, University of Nice–Sophia Antipolis, 1998.

15. R. Carson. *Silent Spring* (Cambridge, Massachusetts: Riverside Press, 1962).

CHAPTER SEVEN

1. René Richard (1904–78): This defender of the environment fought on the Côte d'Azur from 1970 to 1977. His courage and passion in defending his ideas in the field, at a time when concrete was devastating the coast, profoundly impressed me. His battle is illustrated by a pamphlet he co-authored, *La Côte d'Azur assassinée?* (Paris: Roudil, 1971). He founded, with Dr. Alain Bombard and the physicist Louis Leprince-Ringuet, the regional union for the protection of life and nature (URVN), the main organization fighting for the protection of nature in southeastern France.

2. The history of phylloxera (a small sucking bug introduced to France in

1860) described below is a good illustration. Scientific controversies and media polemics on its origin and the gravity of the threat it posed (at the beginning of the invasion), many eradication attempts, involvement of groups of inventors and attention to the problem by committees under the aegis of the Academy of Sciences marked the trajectory of this terrible biological pollutant. See R. Pouget, *Histoire de la lutte contre le phylloxéra de la vigne en France* (History of the Struggle against Phylloxera in France), (Paris: INRA, 1990).

3. See M. de Pracontal, *L'Imposture scientifique en dix leçons* (Scientific fraud in ten lessons) (Paris: La Découverte, 1986).

4. In France alone, *Caulerpa taxifolia* was the subject of more than 560 news articles between 1990 and 1996.

5. To learn more about this concept, I especially recommend a remarkable work: M. Chauvet and L. Olivier, *La Biodiversité, enjeu planétaire* (Biodiversity: The planetary stakes) (Paris: Sang de la Terre, 1993).

6. A human population of 6 billion will be reached or surpassed by the year 2000. For 2025, the estimates are between 8 and 12 billion.

7. On this subject, see "Une Terre en renaissance. Les semences du développement durable," *Savoirs* no. 2, publication of *Le Monde Diplomatique*, 1993.

8. While the term "ecology" is not free of political or partisan connotations, the word "biodiversity" is more neutral. It is difficult to say that someone is a "biodiversicologist."

9. The precautionary principle is echoed in the "responsibility principle," an ethical concept developed by the German philosopher Hans Jonas. See his *The Imperative of Responsibility: In Search of Ethics for the Technological Age* (Chicago: University of Chicago Press, 1984). This work has become a standard reference in ecological circles.

10. M. Serres, *Le Contrat naturel* (The natural contract) (Paris: Francois Bourin, 1992).

11. J. Lovelock, *Gaia: Practical Medicine for the Planet* (New York: Harmony Books, 1991).

12. L. Ferry, *Le Nouvel Ordre écologique* (The new ecological order) (Paris: Grasset, 1992).

13. French specialists prefer to be called "écologues" rather than "écologistes"; the latter has a political connotation.

14. Subsequently, some of the 264 signatories (including 52 Nobel laureates) regretted having fallen into the trap set by those who conceived this aggressive declaration.

344   15. In France, extracts of many reactions have been collected in the journal *Les Cahiers de Global Chance* 1 (1992).

16. See the file on evolution in *Pour la Science*, January, 1997.

17. "The enormity of ignorance and of the task to accomplish" is the title of a chapter of report no. 33 of the Academy of Sciences, entitled *Biodiversité et Environnement* (Paris: Technique et documentation-Lavoisier, 1995).

18. The latest complete inventory is by the American biologist E. O. Wilson, who is also the apostle of the concept of biodiversity. See *The Diversity of Life*, ch. 6, n. 11.

19. Two worldwide programs have been in place since 1992: *Systematics Agenda 2000, charting the life,* championed by scientific systematists' societies; *Diversitas,* espoused by ecologists more oriented toward research on ecosystem function.

20. The program *Natura 2000,* in the framework of the directive "Habitats."

21. This classification is presented in Chauvet and Olivier, *La Biodiversité* (see note 5).

22. See C.-F. Boudouresque et al., *Red book "Gerard Vuignier" of threatened plants, communities, and landscapes of the Mediterranean* (Athens: United Nations Environment Program, 1990).

23. The *"Posidonia* network" was established in 1987 off the coasts of Provence and Côte d'Azur by the governments of this region, under the aegis of GIS-Posidonie.

24. *Waters of the Netherlands: Immediate Action,* Ministry of Transport and Public Works of the Netherlands, 1989; B. Ten Brink et al., "A quantitative method for description and assessment of ecosystems: the AMOEBA approach," *Marine Pollution Bulletin* 23 (1992): 265–70.

25. C. H. Lindroth, *The Faunal Connections between Europe and North America* (New York: Wiley, 1957).

26. See note 2.

27. C. Bright, *Life Out of Bounds* (New York: W. W. Norton, 1998).

28. B. P. Butterfield, W. E. Meshaka, Jr., and C. Guyer, "Nonindigenous reptiles and amphibians," in D. Simberloff, D.C. Schmitz, and T. C. Brown, eds., *Strangers in Paradise,* 123–38.

29. M. Williamson, *Biological Invasions* (London: Chapman & Hall, 1996).

30. Ibid.

31. GESAMP (IMO/FAO/UNESCO-IOC/WMO/WHO/IAEA/UN/ UNEP Joint Group of Experts on the Scientific Aspects of Marine Environmental Protection), *Opportunistic Settlers and the Problem of the Ctenophore* Mnemiopsis leidyi *Invasion in the Black Sea.* Reports and Studies, no. 58 (1997).

32. C. J. Krebs, *Ecology,* 4th ed. (New York: Harper Collins, 1994).

33. D.C. Schmitz, D. Simberloff, R. H. Hofstetter, W. Haller, and D. Sutton, "The ecological impact of nonindigenous plants," in Simberloff, Schmitz, and Brown, eds., *Strangers in Paradise,* 39–61.

34. J. C. McKinley, Jr., "An Amazon weed clogs an African lake," *New York Times,* August 5, 1996.

35. Q. C. B. Cronk and J. L. Fuller, *Plant Invaders* (London: Chapman & Hall, 1995).

36. Ibid.

37. L. E. Johnson and J. T. Carlton, "Post-establishment spread in large-scale invasions: dispersal mechanisms of the zebra mussel *Dreissena polymorpha," Ecology* 77 (1996): 1686–90.

38. U.S. Congress, Office of Technology Assessment, *Harmful Non-Indigenous Species in the United States* (Washington, D.C.: U.S. Government Printing Office, 1993).

39. D. Simberloff, "The biology of invasions," in D. Simberloff, D.C. Schmitz., and T. C. Brown, eds., *Strangers in Paradise,* 3–17.

40. D. Simberloff, pers. comm., 1998.

41. The grandes écoles are prestigious schools at a university level with competitive entrance examinations. They are the training ground for a large fraction of the French bureaucracy and technocracy.

42. See chapter 3, note 47.

43. P. Jouventin, "Why is there a debate on scientific politics in ecology?" *Bulletin d'Écologie* 22 (1991): 253–56; F. Ramade, "Ecological science in France and abroad," *Bulletin d'Écologie* 22 (1991): 257–59; F. Ramade, "Does ecology have a future?" *La Recherche* 253 (1993): 422–26; T. Pilorge, "Debate: Molecular biology against the sciences of nature," *Science et vie,* no. 927 (1994): 48–52.

44. F. Brune, "The violence of ideological publicity," *Le Monde diplomatique,* August 1995, 497.

45. "Holism" comes from the Greek *holos* = entire, indicating the tendency of the universe to construct entities of increasing complexity: inert mat-

346

ter through living matter. Ecology encompasses the study of complex systems in their entirety; its conduct thus has a holistic nature. See F. Ramade, "Does ecology have a future?" (see n. 43).

46. François Ramade goes so far as to claim that to present oneself as having these research goals (taxonomy and ecology of plants) in a university or the CNRS is considered proof of low intelligence if not retardation. See F. Ramade, "Ecological science in France and abroad" (see n. 43).

47. C. Lévêque, "The future of the sciences of nature: contradictory opinions," *La Recherche* 257 (1993): 1013–14.

48. F. Gros, *L'Ingénierie du vivant* (The technology of life) (Paris: Odile Jacob, 1990).

49. In the 30th section of the department of life sciences, entitled, "Biological diversity–populations–ecosystems and evolution."

50. See the remarkable reflection of André Giordan on the study of life: A Giordan, *Comme un poisson rouge dans l'homme* (Like a red fish in man), (Paris: Payot et Rivages, 1995).

51. F. Ramade, "Ecological science in France and abroad" (see n. 43).

52. This threat is evoked in a humorous editorial by the editor-in-chief of one branch of the major journal of the sciences of life. See B. J. Culliton, "Clinical investigation: an endangered science," *Nature-Medicine* 1 (1995): 281.

53. "The beginning scientist no longer has any familiarity with life. Everything has been taught in a theoretical and fragmented mode, . . . a little cell biology, a little physiology, a little immunology, a little endocrinology, and a lot of molecular biology" (A Giordan, *Comme un poisson rouge dans l'homme*, 403).

54. J. T. Carlton, "Neoextinctions of marine invertebrates," *American Zoologist* 33 (1993): 499–509.

55. Among the twenty members of the Academy of Sciences who comprise part of the editorial board of the biology section of *Comptes rendus*, there is a single naturalist, a specialist in freshwater habitats.

56. For all scientific disciplines together, there are some 100,000 journals worldwide. The majority are in English, and they are of highly variable quality.

57. The dissemination of scientific information in the popular media is analyzed well in the book by P. Fayard, *Science aux Quotidiens* (Science in the news), (Nice: Z éditions, 1993).

58. P. Fayard noted "the standard of notoriety of researchers" (*Science aux Quotidiens*, 77). See also S. Erkman, "Influencing of scientific reporters," *Le Monde Diplomatique*, October 1996.

59. According to P. Fayard (*Science aux Quotidiens*, 115), this circulation and sale of scientific information is beginning to generate acid comments by scientific reporters who deplore the English-speaking hegemony in the production and recognition of scientific scoops.

60. The editors of primary journals do not always look kindly on these new modes of scientific communication. The most prestigious journals require researchers not to announce a result before the article is published and refuse to publish results already circulating on the Internet. Scientific publication on paper will truly be threatened when the pages of the Internet can be sanctioned by a rapid and universally recognized system of refereeing.

61. Since the advent of the law on orientation and programming of research and development (1982), the researcher has three official functions: develop his or her discoveries, enhance their prestige, and disseminate them to the public at large.

62. G. Delacôte and R. Barbault, "Synthesis of activities of the workshop: The politics of communication of results of environmental research," *Lettre du Programme environnement* (CNRS) 13 (1994): 35.

63. The National Committee on Ethics in the Life and Health Sciences recently expressed concern at the pathways of information in medicine and biology: "The practice whereby researchers go directly to the popular media to inform them of a result, preventing the crucial preliminary evaluation by one's peers, must be generally deplored" (opinion of the CCNE, July 1995).

64. "On scientific communication," Reflections of the Committee on Ethics in the Sciences, in response to a letter submitted by François Fillon, CNRS, Communication to the Department of Life Sciences, January 1995.

65. A. Ghazi, "Mr. Benveniste, science, and *Le Monde*," *Le Monde*, May 29, 1996.

66. J. Baudrillard, "Information at the meteorological level," *Libération*, September 18, 1995.

67. M. de Pracontal observes that "the universe of the popular media is the Eldorado of imposters" (*L'Imposture scientifique en dix leçons*, 122).

68. "Homeopathy. 1995 update on the evaluation file," *Prescrire* 15 (1995): 674–84.

69. See B. Latour, *La Science en action* (Science in action) (Paris: Gallimard, 1995), 84.

70. See G. Delacôte, "Environment, research, and society," *Lettre du Programme environnement* (CNRS) 13 (1994): 50.

71. B. Latour, *La Science en action.*

72. These distinctions were analyzed in a memoir devoted to the analysis of the popularization of the *Caulerpa taxifolia* case: M. Fouchard, "*Caulerpa taxifolia* in *Le Monde* and *Nice-Matin:* from a scientific polemic to the emergence of a new function, the critique of science," Master of Science in information and communication, University of Nice-Sophia Antipolis, 1994.

73. J. Clottes and J. Courtin, *La Grotte Cosquer* (Paris: Seuil, 1994); *The Cave beneath the Sea: Paleolithic Images at Cosquer* (New York: H. N. Abrams, 1996).

74. The cold fusion affair was analyzed by a sociologist who compared the production of scientific articles (the controversy) to that of popular articles (the polemic). B. V. Lewenstein, "The saga of cold fusion," *La Recherche* 266 (1994): 636–40.

75. "The manipulation by some laboratories of networks of influence is not in itself very different from the maneuvers perpetrated by dubious and corrupt politicians." (D. Bougnoux, "Science threatened by the media," *Le Monde Diplomatique,* September 1995).

76. W. Broad and N. Wade, *Betrayers of the Truth* (New York: Simon and Schuster, 1982).

77. Suren Erkman, "The role and practice of scientific journalism," European Master's degree in science, society, and technology, École polytechnique Fédérale de Lausanne, Switzerland, 1995. See also "Influencing of scientific reporters," *Le Monde Diplomatique,* October 1996.

78. Martine Barrère, scientific journalist, said on this point, "It is necessary to construct another science based on a new relationship between scientists and citizens." (See her article "Science and society: What do they share?," *Les cahiers de Global Chance* 6 (1996): 5.

79. M. Massenet, *La Transmission administrative du SIDA. Qui sont les vrais responsables?* (The administrative transmission of AIDS. Who is really responsible?) (Paris: Albin Michel, 1992).

1. See A. Meinesz and B. Hesse, "Introduction and invasion of the tropical alga *Caulerpa taxifolia* in the northwest Mediterranean," *Oceanologica Acta,* 14, no. 4 (1991): 415–26.

2. For a summary in lay language of *Caulerpa* structure and biology, see W. P. Jacobs, "Caulerpa," *Scientific American* 271, no. 6 (December 1994): 100–105.

Index